Surveillance Schools

DOI: 10.1057/9781137308863

Crime Prevention and Security Management

Series Editor: **Martin Gill**

Titles include:

Paul Almond
CORPORATE MANSLAUGHTER AND REGULATORY REFORM

Rachel Armitage
CRIME PREVENTION THROUGH HOUSING DESIGN
Policy and Practice

Joshua Bamfield
SHOPPING AND CRIME

Daniel Donnelly
MUNICIPAL POLICING IN THE EUROPEAN UNION
Comparative Perspectives

Paul Ekblom
CRIME PREVENTION, SECURITY AND COMMUNITY SAFETY USING THE 5IS FRAMEWORK

Adam Graycar and Tim Prenzler
UNDERSTANDING AND PREVENTING CORRUPTION

Bob Hoogenboom
THE GOVERNANCE OF POLICING AND SECURITY
Ironies, Myths and Paradoxes

Daniel McCarthy
'SOFT' POLICING
The Collaborative Control of Anti-Social Behaviour

Tim Prenzler
POLICING AND SECURITY IN PRACTICE
Challenges and Achievements

Emmeline Taylor
SURVEILLANCE SCHOOLS
Security, Discipline and Control in Contemporary Education

Jan van Dijk, Andromachi Tseloni and Graham Farrell (*editors*)
THE INTERNATIONAL CRIME DROP
New Directions in Research

Adam White
THE POLITICS OF PRIVATE SECURITY
Regulation, Reform and Re-Legitimation

Crime Prevention and Security Management
Series Standing Order ISBN 978-0-230-01355-1 **hardback**
978-0-230-01356-8 **paperback**
(*outside North America only*)

You can receive future titles in this series as they are published by placing a standing order. Please contact your bookseller or, in case of difficulty, write to us at the address below with your name and address, the title of the series and one of the ISBNs quoted above.

Customer Services Department, Macmillan Distribution Ltd, Houndmills, Basingstoke, Hampshire RG21 6XS, England

---

palgrave pivot

Surveillance Schools: Security, Discipline and Control in Contemporary Education

Emmeline Taylor
Senior Lecturer, Australian National University

palgrave macmillan

DOI: 10.1057/9781137308863

First published 2013 by
PALGRAVE MACMILLAN

Palgrave Macmillan in the UK is an imprint of Macmillan Publishers Limited, registered in England, company number 785998, of Houndmills, Basingstoke, Hampshire RG21 6XS.

Palgrave Macmillan in the US is a division of St Martin's Press LLC, 175 Fifth Avenue, New York, NY 10010.

Palgrave Macmillan is the global academic imprint of the above companies and has companies and representatives throughout the world.

Palgrave® and Macmillan® are registered trademarks in the United States, the United Kingdom, Europe and other countries

ISBN: 978-1-137-30887-0 EPUB
ISBN: 978-1-137-30886-3 PDF
ISBN: 978-1-137-30885-6 Hardback

This book is printed on paper suitable for recycling and made from fully managed and sustained forest sources. Logging, pulping and manufacturing processes are expected to conform to the environmental regulations of the country of origin.

A catalogue record for this book is available from the British Library.

A catalog record for this book is available from the Library of Congress.

www.palgrave.com/pivot

DOI: 10.1057/9781137308863

Contents

DOI: 10.1057/9781137308863

Preface

In the early 2000s, I embarked upon a PhD in Manchester, United Kingdom that aimed to explore the societal impacts of CCTV. The UK at that time was undergoing an aggressive, state-funded roll out of CCTV systems that would see coverage across England and Wales rocket from a mere 100 public surveillance cameras in 1990 to over a million in 20 years. The UK was fast emerging as the 'CCTV capital of the world' but there wasn't a huge amount of empirical research being conducted on the impact it was having on society, other than a handful of evaluations that specifically focused on whether CCTV 'worked' in reducing crime. Largely omitted from the agenda was any discussion about the 'side effects' of CCTV and how it might impact on social interaction, privacy and civil liberties.

As with many PhD scholars, the task ahead seemed daunting and the many facets of CCTV, and more broadly surveillance, was unwieldy. One day I sat talking with my sister, ten years my junior. I told her that I was exploring the use of CCTV in more depth, and her response is something that has shaped my scholarly enquiry ever since: 'will you be looking at the CCTV in my school?' I thought I had misheard her at first. Cameras in her school? She was 11 years old. It became apparent that a school just up the road from where I was living, in a lower-middle class suburban area, had installed a very sophisticated CCTV system with cameras in the classrooms, corridors and the pupil's toilets. There was extensive, near-seamless coverage that any urban metropolis or police department would be envious of. In fact this one school with little over

DOI: 10.1057/9781137308863

1000 pupils had more CCTV cameras than the city centre public scheme at the time. I couldn't understand it, and immediately I had a research topic that would continue to fascinate me a decade later. Over the years, I have had many similar encounters talking to pupils as they somewhat nonchalantly tell me about the new fingerprinting systems, the random searches without suspicion, and the omnipresent CCTV cameras that have become a routine part of their school day.

My incredulity is less with the technologies and processes themselves, but more the lack of consideration that has been given to their use in schools. Blind faith in technology as a quick-fix panacea, draconian zero tolerance strategies, a generalised mistrust of youth and the infiltration of corporate values into education have taken us down a path of creating Surveillance Schools that are not necessarily safer, better or more conducive to learning. These policies and technologies are too often implemented at an exorbitant cost to under-resourced schools and premised on 'commonsense' presumptions of efficacy. My concern lies with the impact this is having on pupils, teachers and staff, and more broadly society. In the school environment, pupils are taught to accept 'the way things are', particularly when activities are embedded into the pedagogical apparatus. Pupils have a relative lack of power in an institution that is oriented around an axis of control and discipline. Their dual casting, as potential perpetrators and vulnerable victims, make them ideal candidates for invasive surveillance practices.

But what can we make of these developments in schools? *Surveillance Schools* maps out the trends in technological surveillance and begins to unpick some of the causal factors driving this revolutionary change in educational practice.

DOI: 10.1057/9781137308863

Acknowledgements

This book would not have been possible without the support and influence of numerous individuals to whom I am indebted. Throughout my career as a researcher I have engaged with many colleagues and friends who have helped me to generate new ways of thinking about surveillance, its impact and the role of schools in society.

Several people read sections of the manuscript and provided excellent feedback. I would like to thank my colleagues at the Australian National University for engaging with my research and ideas, and my former colleague, Professor Martin Gill, for his ongoing encouragement.

Many teachers, staff and pupils have shaped my understanding of Surveillance Schools. I am very grateful for the forthright views and experiences that they shared with me, particularly those who participated in the case studies in the North of England.

I would like to thank the team at Palgrave Macmillan for making this book a reality.

On a personal note, I wish to thank my family and friends for their support. Jamie provided encouragement and a sense of perspective when it was most needed and I couldn't have done it without him.

DOI: 10.1057/9781137308863

Series Editor's Introduction

This book provides a powerful reminder of why surveillance merits careful attention and why this takes on a special significance in a school setting. While surveillance can provide for safer and more secure settings, claims on the extent to which these can be assured by different types of surveillance technologies and practices can be overstated, and the negative consequences can completely nullify any positive effects. These issues are a prime concern of this book in which Emmeline Taylor analyses the range of ways in which teachers and pupils are watched over, and how these are used alongside other strategies, such as security screening, personal searches and zero tolerance of some types of behaviour to instil a negative culture to the environment in which children are educated.

Via her own work on CCTV in schools, she highlights the ways in which children felt it 'undermines privacy' and contributes to a 'manifestation of mistrust', having a 'chilling effect' on creativity, while simultaneously initiating a process of 'distanciation' (denying the opportunity for social interaction). Emmeline draws from a range of theories and writers to explain the consequences this has on both generating a false sense of safety and security but more importantly, for her interests, in leading to the isolation and marginalisation of certain groups of young people who are caught up in overly punitive environments. This book does not just point to the danger of civil liberty intrusions; it shows the ways in which this is happening. In response she proposes a new way of responding to indiscipline in schools, one that moves away from a punitive approach, namely restorative justice. She also argues the case for surveillance studies to be established on the school curriculum to enable young people to understand the dangers posed to rights and freedoms but also to enable pupils to engage with surveillance meaningfully. She sets out the aim of ensuring young people are 'capable of contributing to a society based on democratic and inclusive principles' while 'nurturing critical thought'.

In this critical account of the dangers of surveillance technologies Emmeline stops short of rejecting them altogether; they have a place when properly specified, deployed and managed and set in a culture that recognises the dangers and seeks to respond to them. No doubt this book will court controversy, and so it should. While the dangers of not prioritising the safety and security of our children are very real indeed, these have been widely recognised, the consequences of too much surveillance much less so. This book seeks to redress the balance.

Martin Gill
September 2013

DOI: 10.1057/9781137308863

Introduction

Abstract: *Surveillance Schools are emerging around the globe characterised by new technologies and practices that identify, verify, categorise and track pupils in ways never before thought possible. The ramifications are huge. We are stood on the verge of what could be one of the most colossal shifts in societal structure and organisation as Surveillance Schools normalise new structures of power and control.*

Taylor, Emmeline. *Surveillance Schools: Security, Discipline and Control in Contemporary Education.* Basingstoke: Palgrave Macmillan, 2013.
DOI: 10.1057/9781137308863.

It's Monday morning, Chloe removes her watch and walks through the metal detector archway. Having cleared the routine check, she proceeds past the uniformed police officer to the fingerprint scanner where she places her finger onto the reader until she hears a beep signalling a match. She walks down the camera-lined corridor, not only visually monitored but also remotely tracked by the microchip that is sewn under the emblem on her jacket. Entering the bathrooms, she ensures that the cubicle door is firmly closed and using the hook on the back of the door she attempts, as best as she can, to position her coat across the small crevice between the door and frame to ensure that the dome CCTV camera mounted high up on the wall cannot see within. During the day, Chloe will be captured on a surveillance system comprised of more than 100 cameras, everything she purchases to eat will be recorded in a database and compiled into a weekly report, and her every movement will be tracked via a radio frequency identification system. Chloe is not in a high security prison, nor is this the opening to a futuristic sci-fi novel. Chloe is at school, and this is now.

I developed a similar fictional account to the one above for a short article that was published in *Living Ethics* magazine in August 2012. I wanted to convey, in a limited space, the assemblage of surveillance technologies that were appearing in schools and communicate the intensity with which they monitor pupils. My intention was to generate empathy with the experiences of young people as subjects in an institution increasingly organised around and governed by surveillance principles and practices, both structural and technological.

Dramatised scenarios aside, Surveillance Schools are emerging around the globe characterised by an array of routine practices that identify, verify, categorise and track pupils. Visual surveillance technologies such as Closed Circuit Television (CCTV) are common, and a range of biometric technologies such as fingerprinting, iris scanning and palm vein readers are finding their way on to campus. Radio Frequency Identification (RFID) microchips are being embedded in school ID cards and 'smart uniforms' to monitor pupil's movements on and off the school premises. Globally, many schools require that pupils pass through security arches upon entering the school premises or use hand-held metal detector wands to perform random searches without requiring consent, or even suspicion. The array of devices, whilst certainly not present in all schools, represent a shift from human-centred strategies of discipline to technological mechanisms that bring with them new narratives of risk, fear and control. However, Surveillance Schools are not just comprised of technological apparatus; many have full-time uniformed police officers, armed in some countries; patrolling the corridors and classrooms, they subscribe to routine drug testing and use sniffer dogs to search students and their possessions stored in transparent lockers and bags. Analysed within a context of zero tolerance initiatives, record suspension levels and fortified school buildings, one has to consider the impact that this is having on children.

It can be argued that schools have always had surveillance at their core; registration confirms attendance, student reports compound activity, continual examination and assessment monitors progress, the containment of pupils on a bounded campus enables close observation of behaviour, and the contravention of rules attracts swift and often visible punishment. Schools as sites of surveillance are nothing new. Indeed many school architects embed the need for surveillance in their designs.[1] However, a raft of new surveillance technologies and practices have created educational institutions in which pupils are emerging as the most heavily surveilled population in countries such as the United Kingdom and United States.[2] This book explores the rise of the Surveillance School, mapping the driving forces behind them, and explicating the impact that these radical changes in educational institutions are having on pupils, teachers, schools and, more broadly, society.

What is surveillance?

The word 'surveillance' has its origins in the French verb 'surveiller', which literally translates to 'watch over'.[3] Applying this literal interpretation to young people, a child may be 'watched over' because one is concerned for their safety: parents watching their children playing in the park, for example. Idioms such as *keeping an eye on the kids* are commonly used, and on the one hand can refer to ensuring that young people do not cause trouble or mischief, or on the other, it can be perceived as a protective measure to stave off exposure to potential dangers. Watching can be used to ensure compliance with rules and regulations, such as the invigilation of exams. Whereas some young people may be watched over because their appearance is somehow categorised as suspicious or their

DOI: 10.1057/9781137308863

behaviour has been defined as undesirable or deviant. They may attract attention, directly or indirectly, because of their class status, gender, religion or race. There are of course sensitivities around watching and documenting the activities of children as captured in blanket bans on photography in schools, including activities that parents typically attend such as sports day and performances. The nuanced nature of surveillance comes to the fore when these same schools that prohibit parents taking photographs or video recording a school production, routinely record the movements, habits and behaviours of pupils via sophisticated surveillance technologies.

More formally, surveillance has been defined by Anthony Giddens as the 'control of information and the superintendence of the activities of some groups by others'.[4] A similar definition has been proposed by David Lyon, who describes surveillance as 'the focused, systematic and routine attention to personal details for the purposes of influence, management, protection or direction'.[5] A more technological definition has been offered by the European Union Directorate General for Research (EUDGR), which proposes that surveillance can be defined as 'devices or systems which can monitor, track and assess movements of individuals, their property and other assets'.[6] From these definitions, and there are many more, we can see that there is a power dynamic inherent in any surveillance mechanism, whereby one group possesses the means and power to oversee, direct, control, even manipulate, the activities of others. Surveillance practices are not equal in their nature, and information processing rarely takes place on a level playing field.

The term 'surveillance' upon first account conjures certain images, such as covert police operations or spies monitoring and documenting the activities of their targets. Admittedly it is perhaps more appealing to conceive surveillance as furtively conspiratorial and clandestine, but in reality relatively little surveillance operates in this way. Surveillance is in fact a very 'normal' by-product of the modern day capitalist society. The production of vast amounts of data, and its processing by various entities, is the consequence of contemporary living, generated by every day, mundane and banal activities such as shopping, driving, emailing, social networking, using a cell phone, working and learning. We are forever entangled with technological surveillance as every digitally mediated action sheds information about us. It is true that surveillance often has a hidden side; increasingly it is not so much about 'seeing' people, but rather interacting with and processing characteristics of their 'data

DOI: 10.1057/9781137308863

bodies',[7] or 'data doubles'.[8] As such, it is not always apparent when the surveillant apparatus is in full throttle, processing and amalgamating disparate details of our lives and facilitating decisions of entitlement and eligibility based on our abstracted profiles. In the surveillance age virtual representations of the self have become invaluable and profitable commodities as well as units of analysis.

Dystopian visions of surveillance

Many scholars concerned with surveillance quickly find themselves on the well-trodden path that ultimately leads to urban dystopia or totalitarian regime, highlighting the negatives of surveillance whilst denying any potential benefits. Surveillance is perceived as an imposition on unsuspecting and passive citizens, but this fails to recognise that often individuals are complicit in their own surveillance and at times even court the salacious opportunity for exposure that various forms of surveillance provide. In fact, none of the fundamental components of society that the average citizen in a capitalist nation has come to expect and enjoy would be possible without surveillance. That is not to say that surveillance technologies operate without significant consequence or that they can be treated as something benign, but rather a Luddite-style rebuttal of new technologies is flawed and unviable. Surveillance is more complicated than that.

Surveillance is complex and messy. It can be positive and enabling, liberating and rewarding whilst at the same time oppressive and dictatorial. Surveillance practices underpin the essence of democracy by allowing the administration of the voting system; transactions are monitored by financial institutions to detect fraud; loyalty programmes issue us with rewards for our custom and collate data on our spending habits to target campaigns that align with our preferences; visual surveillance can help identify some offenders through CCTV footage, whereas GPS can assist the emergency services to locate individuals in need, and dataveillance provides us with entitlements to benefits and welfare. However, illustrating the inherent ambiguity of surveillance, it is integral to totalitarianism and can undermine privacy and freedom of expression.[9] Surveillance practices are predisposed to sort people into typologies based on abstracted profiles, which ultimately attracts discrimination. It denies some people access to services based on predefined categories of risk and suspicion, and is often unfairly focused upon certain groups, especially along the lines of age, race and gender. It is not surprising

DOI: 10.1057/9781137308863

given the complexities of surveillance that David Lyon describes it as having 'two faces'.[10] This should not be interpreted as duplicity, but rather that any surveillance mechanism is characterised by the coexistence of antagonistic elements.

The understanding of surveillance is complicated further by people's reaction to it. Some individuals resign themselves to the 'necessity' of surveillance in the post-9/11 world, often subscribing to the vacuous and prosaic idiom 'if you have nothing to hide you have nothing to fear', and go about their daily business seemingly unaffected. Others are willing to forego privacy in exchange for the protection they perceive surveillance offers them, as if liberty and safety existed in a zero-sum model. Further still, viewed through a neoliberal lens, some believe that surveillance has a role in bringing market principles to the education system and enables corporate-driven efficiency savings.

Technological determinism

In light of the above, both advocates and critical scholars of surveillance must be cautious that they do not fall into the trap of technological determinism; 'an unquestioning belief in the power of technology, whether benign or malevolent'.[11] Overstating capabilities, whether good or bad, can only ever stultify understanding and result in poor policy decision-making and regulation. Technologies are themselves social entities[12] that cannot be divorced from the social world in which they are conceived and applied. They are entwined with ideology, power, institutional objectives and social inequality. Surveillance practices alter behaviour; indeed they are intended to do so, and so their existence can never be viewed as impartial. Since technologies cannot be divorced from the societies in which they are created and proffer, neutrality should never be presumed. As Martin Heidegger warned:

> Everywhere we remain unfree and chained to technology, whether we passionately affirm or deny it. But we are delivered over to it in the worst possible way when we regard it as something neutral; for this conception of it, to which today we particularly like to do homage, makes us utterly blind to the essence of technology.[13]

This is particularly salient for the contemporary school upon which a host of forces play. Largely owing to its multiple forms, differing objectives and motivations for its use, surveillance is irrevocably abstruse and no one reading can resolve the infinite complexities that it brings to

DOI: 10.1057/9781137308863

societal structures on miso, micro and macro levels. There is perhaps just one thing about surveillance that is not steeped uncertainty, the fact that society is increasingly being shaped by its manifestations and practices.

The surveillance society

Surveillance is not a new phenomenon. Early incarnations of routine surveillance can be located in the formation of the nation state which necessitated a means of keeping track of citizens for activities such as gathering taxes, distributing entitlements or mobilising an army.[14] Although Western societies prior to the modern era documented some details of their populations to facilitate and coordinate their activities, it is only in the past 30 years that an unprecedented upsurge in 'everyday' surveillance activities has arisen. Amidst political and economic changes, the revolutionary advances in technology has driven and responded to the desire for surveillance in developed capitalist societies – on the one hand seeking out new markets for technological developments, and on the other responding to an increasing demand to identify, verify, sort and categorise individuals. It is not possible to isolate one sole trend or commanding force; a constellation of phenomena have come together to create the Surveillance School.

Following the September 11 terrorist attacks on the World Trade Centre in New York in 2001, surveillance practices have intensified in all facets of daily life. As fear became amplified, objection to the swell of surveillance practices and regimes quietened.[15] Concerns about civil liberties have been relaxed as the demand for protection *from* the state has been replaced by the demand for protection *by* the state.[16] In response to growing global fears the public has become, or is at least represented as being, risk-averse, and intently focused on protection and preservation over all other concerns. As David Garland notes, 'the risk of unrestrained state authorities, of arbitrary power and the violation of civil liberties seem no longer to figure so prominently in public concern'.[17] Social, political, economic and corporate concerns have manifested themselves in the surveillance of everyday practices and the hyper-securitisation of environments, including schools. Surveillance becomes 'necessary' to control crime and criminals. Activities, behaviours and institutions have become imbued with new narratives of risk, and these serve not just to securitise them, but to open them up to global corporate forces.

DOI: 10.1057/9781137308863

Moral panic about endemic school violence has provided a veneer of acceptability, even reverence, for the increased surveillance of pupils and their environment. Idiosyncratic events, such as school shootings or knife crimes are transmitted to a global audience as evidence that schools are dangerous territories that require intensive control. More stringent security is presented as the only option in an arms race against disenfranchised youth and predatory loners, and of course surveillance security vendors are always on hand to point out how their product can prevent further atrocity.

Surveillance schools: genesis and impact

There are multifarious drivers of surveillance in schools, and understandably these vary over time, by country, by technology, objectives, location and so on. *Surveillance Schools* aims to cast light on the developments and provide insight and understanding as to how and why these processes are taking place. Monitored like prisoners, chipped like cattle? Is this the future for pupils around the globe? Or are surveillance technologies simply automating processes that have always been and always will be a part of the school day? With these questions in mind, the text aims to

1. outline current trends in surveillance in schools;
2. explore the impact of new surveillance technologies on pupils, teachers and society;
3. identify the key factors driving surveillance practices.

It is perhaps expedient at this point to outline the parameters and limitations of this book. First and foremost, the book aims to provide an overview of trends in Surveillance Schools, drawing upon examples from around the world to illustrate how surveillance has promulgated and manifested itself in different contexts. However, the analysis is propelled by the processes taking place in the developed capitalist western world. The book oscillates between being a 'zoomed out' abstracted account of trends and processes, and providing instances of a more 'zoomed in' nature to contextualise and articulate specific examples (largely drawn from England and North America). Attempting to navigate the multiple and varied factors that shape surveillance in schools in a limited space will of course result in significant omissions, simplification and conversely, the overstatement of some examples to foreground their significance.

DOI: 10.1057/9781137308863

The trends identified are certainly not represented in equal measure in all countries that are witnessing the rise of Surveillance Schools, and of course local context is paramount. It is important to recognise that trends are often nuanced and idiosyncratic: a précised distillation of them can never do the details justice, but my intention is to try and map some of the key structural and generalised facets of Surveillance Schools.

Chapter 1 explores the confluence of surveillance practices in schools including Automatic Fingerprint Identification Systems (AFIS), CCTV and Radio Frequency Identification systems (RFID). Largely descriptive, the chapter provides an overview of new systems of surveillance, their objectives and outcomes. Set within the context of structural surveillance practices such as the increasing use of permanent onsite police officers, high stakes examination, zero tolerance approaches to discipline, random searches and drug testing, schools are possibly one of the most invasive and inimical environments for young people. Chapter 2 offers an empirical case study on the use of CCTV in three secondary schools in the North of England to explicate the way pupils experience CCTV at their school.

Drawing upon a number of empirical studies, Chapter 3 examines the impact that the proliferation of surveillance in schools is having on pupils and, more broadly, on society. Surveillance transforms behaviour and interaction, as well as undermining societal qualities such as privacy and trust. By embedding surveillance into pedagogical apparatus, young people are being habituated to unprecedented levels of scrutiny and control without being equipped with the knowledge and tools required to navigate the digital world. It is not all doom and gloom; pockets of resistance do exist, and sometimes take on rather surprising forms.

Chapter 4 examines the role of discipline in the school before exploring the conflux of socio-political, cultural and economic factors that have contributed towards the ascension of the Surveillance School. The chapter explores a world in which welfare provision is being rolled back, growing proportions of the population are incarcerated, and where pupils are increasingly educated in criminalised spaces. Big Brother is ultimately big business and alongside criminalisation, the corporatisation of schools is found to be a significant catalyst in the genesis of Surveillance Schools.

The conclusion draws together the disparate lines of argument developed throughout the text to consider the future of surveillance in schools and the impact this will have on society. Several suggestions are advanced

DOI: 10.1057/9781137308863

on how to wrestle the school from corporate concerns and reinvigorate democratic values, justice and citizenship. Restorative Justice is drawn upon as a possible alternative to the ethos of zero tolerance to refocus the school system towards reparative rather than punitive approaches to discipline.

Why research surveillance in schools?

Schools hold a unique position in society as they are largely responsible for the education and socialisation of new members of society. In this capacity they have the power to shape the way young people experience and negotiate surveillance apparatus and practices. The school has a number of features which make it a particularly pertinent unit of analysis.

Socialisation and discipline

School has always been about something more than education; its business is also socialisation and moral instruction. Social commentators have long recognised the role of the school in instilling moral order and discipline in pupils. For some, such as Emile Durkheim, education provides a common morality and disciplinary instruction that maintains social solidarity and cohesion. For others, the disciplinary regimen of the school has historically facilitated the smooth transition of the working classes into the workplace.[18] Conformity and submission are key elements of this process and are identifiable in numerous mechanisms and rituals associated with the school: the school uniform depersonalises and embodies compliance, the school bell signals when to start things and when to end, the timetable regulates the passing of time and the subject offerings reinforce the notion that pupils are passive recipients of others' ontology. By locating surveillance practices in schools they become embedded in the pedagogical apparatus, part of the daily insipid routine, and thereby normalised and accepted by young people. It becomes ordinary, expected and even natural.

A consistent and routine population

The school represents a unique environment that sets it apart from other contexts. For one, attendance is usually obligatory and often underpinned

DOI: 10.1057/9781137308863

by prevailing laws. The majority of young people are required to attend classes and submit to their behaviour being governed whilst on campus. This is in sharp contrast to the relatively transient populations that are subjected to, say, visual surveillance in town centres, car parks and public transport for example. Those who experience surveillance in a school are 'semi-captive' and the fact that the same individuals inhabit the same space on a daily basis means that surveillance forms part of their lived environment, as commonplace and mundane as the blackboard at the front of the classroom. The closed nature of the institution is an important characteristic and has implications for numerous theories relating to surveillance, such as the superlative rumination that has been lavished upon the Panopticon (discussed in detail in Chapter 4).

Watchers and the watched: the dynamic of control

The legitimacy of surveillance is often premised on a shared understanding of who, what, where and for what reason an individual or group is a valid target of surveillance. Typically surveillance discourse demarcates between 'us', the law-abiding citizen, and 'them', the deviants, criminals, and terrorists, who are of course the intended subjects of techno-security surveillance. In this binary conception, support for surveillance is based on the belief that surveillance is solely aimed at controlling 'them' whilst providing protection for 'us'. However, in institutions such as the school this belief becomes markedly less intuitive. Students are often found to be the primary subjects of observation, and seldom due to any deviant intent or suspicion. Teachers and staff are also snared by the surveillance web in an institution increasingly characterised by high stakes standardised tests, scripted lessons and CCTV with audio capability to monitor teacher performance. Teachers occupy a dichotomous position of both 'watched' and 'watchers', the 'watched watchers'. However, as discipline is outsourced to police and private security and technologies such as fingerprinting and CCTV are applied to them, they are increasingly validated as subjects of the surveillant gaze. Whilst some teachers have welcomed this in what they perceive to be a growing litigation age, others have lamented a 'chilling effect' descending upon classrooms.

Even the school itself cannot move beyond surveillance. School league tables readily accessed via the Internet place schools under scrutiny and drive renewed efforts to privatise education.[19] The dynamic of the school is being reconfigured by surveillance and this has important ramifications for education and society.

DOI: 10.1057/9781137308863

Young people as the focus of surveillance

Children's voices are seldom heard in surveillance studies and until recently they received relatively little consideration. A study of CCTV in a British city found that teenagers were targeted by camera operators 'for no apparent reason', whereas in contrast, 'those over thirty years old are rarely the subject of surveillance'.[20] This highlights the need to understand the subjectivities that lead to their surveillance. More broadly, research on schools has tended to ignore the perceptions of pupils and teachers, but rather it has been based upon '"what we all know about school" as, more or less, informed outsiders'.[21]

A recent swell in research exploring the surveillance of children and childhood has emerged and a number of scholars have focused upon facets of surveillance in educational institutions. In *Selling Us the Fortress*, Ronnie Casella casts light on the increasing connections between the security industry and US public schools, and Kenneth J Saltman and David A Gabbard echo the sentiment of the corporatisation of schools, adding a dimension of militarisation in their edited collection *Education as Enforcement*. Aaron Kupchik's *Homeroom Security* focuses on the transformation of disciplinary mechanisms in contemporary schools , whereas Kathleen Nolan explores the growing trend for full time onsite police personnel. Understanding the surveillance of young people is paramount at a time when the status of childhood and youth is being challenged. As Henry A Giroux states, 'increasingly children seem to have no standing in the public sphere as citizens'.[22]

I will argue throughout that far from simply automating existing processes, Surveillance Schools are at once a manifestation of the society in which we now live that is increasingly premised on containment and control, and a key pivotal driver that will reconfigure interpersonal relationships, interactions, and expectations around privacy and civil liberties in the future. Schools are perhaps the most significant consumers of surveillance technologies, and not just on account of their demonstrable appetite. Their unique position in society bestows upon them the ability to determine how future generations will understand, perceive and experience surveillance. Taking schools as microcosms of society, they can provide a prophetic glimpse of the surveillance vista of the future, permitting insight into how the micropolitics of social control operate on an everyday basis.

DOI: 10.1057/9781137308863

Notes

1 Markus, T.A. (1993) *Buildings and Power*. London: Routledge.
2 Taylor, E. (2012) 'The rise of the surveillance school', in K. Ball, K.D. Haggerty, and D. Lyon (eds), *Handbook of Surveillance Studies*. Oxon: Routledge, 225–231.
3 Lyon, D. (2007) 'The search for surveillance theories', in D. Lyon (ed.), *Surveillance Studies: An Overview*. Cambridge: Polity Press, 3–20, p. 13.
4 Giddens, A. (1985) *A Contemporary Critique of Historical Materialism: The Nation-state and Violence*, vol. 2. Cambridge: Polity, p. 2.
5 Lyon, D. (2007).
6 European Union Directorate General for Research (1998) *An Appraisal For Technology of Political Control – Report*. Brussels: EUDGR.
7 Stalder, F. (2002) 'Opinion: Privacy is not the antidote to surveillance', *Surveillance & Society* 1 (1): 120–24.
8 Haggerty, K.D. and Ericson, R.V. (2000) 'The surveillant assemblage', *British Journal of Sociology* 51 (4): 605–22.
9 Giddens, A. (1985).
10 Lyon, D. (1994) *The Electronic Eye*. Cambridge: Polity Press, p. 33.
11 Norris, C. and Armstrong, G. (1999) *The Maximum Surveillance Society: The Rise of CCTV*. Oxford: Berg, p. 9.
12 Monahan, T. (2006) 'Preface', in T. Monahan (ed.), *Surveillance and Security: Technological Politics and Power in Everyday Life*. New York: Routledge, ix–xi.
13 Heidegger, M. (1977) *The Question Concerning Technology and Other Essays*. London: Harper and Row, p. 4.
14 Giddens, A. (1985); Lyon, D. (1994).
15 Lyon, D. (2007) 'The search for surveillance theories'; Lyon, D. (2007) 'Resisting surveillance', in S.P. Hier and J. Greenberg (eds), *The Surveillance Studies Reader*. Berkshire: Open University Press, 368–77; Bloss, W. (2007) 'Escalating U.S. police surveillance after 9/11: an examination of causes and effects', *Surveillance & Society* 4 (3): 208–28.
16 Garland, D. (2001) *The Culture of Control: Crime and Social Order in Contemporary Society*. Chicago: University of Chicago Press.
17 Ibid., p. 12.
18 Bowles, S. and Gintis, H. (1976) *Schooling in Capitalist America: Educational Reform and the Contradictions of Economic Life*. New York: Basic Books; Foucault, M. (1977) *Discipline and Punish: The Birth of a Prison*. London: Penguin Books.
19 Hope, A. (2005) 'Panopticonism, play and the resistance of surveillance: case studies of the observation of student Internet use in UK schools', *British Journal of Sociology of Education* 26 (3): 359–73.

DOI: 10.1057/9781137308863

20 Norris, C. and Armstrong, G. (1999).
21 Ball, S.J. (1987) *The Micro-Politics of the School*, 2nd edn. London: Methuen & Co. Ltd, p. 7.
22 Giroux, H.A. (2003) 'Racial injustice and disposable youth in the age of zero tolerance', *International Journal of Qualitative Studies in Education* 16 (4): 553–65, p. 554.

DOI: 10.1057/9781137308863

1

Surveillance Schools: A New Era in Education

Abstract: *The school gates have been opened to a variety of surveillance technologies including CCTV, metal detectors, fingerprinting, online monitoring, facial recognition and palm vein scanners. Many have full time uniformed police officers patrolling the corridors and classrooms, they subscribe to random drug testing and use sniffer dogs to search students and their possessions stored in transparent lockers and bags. Analysed within a context of zero tolerance initiatives and fortified school campuses, one has to consider the impact that this is having on children.*

Taylor, Emmeline. *Surveillance Schools: Security, Discipline and Control in Contemporary Education.* Basingstoke: Palgrave Macmillan, 2013.
DOI: 10.1057/9781137308863.

Closed circuit television (CCTV)

Visual surveillance devices such as closed circuit television (CCTV) currently represent the most common manifestation of surveillance technology in schools internationally. In the UK it is estimated that 85 to 90 per cent of schools have CCTV, whereas in the US a 2009–2010 survey found that 84 per cent of high schools, 73 per cent of middle schools and 51 per cent of primary schools used security cameras.[1] Asian countries such as the Philippines, China and South Korea are rapidly expanding the use of CCTV in schools, as is Australia. Given the exponential rise in CCTV across the urban landscape,[2] it is perhaps not surprising that it has thrived in schools.

In September 2012 Big Brother Watch, a British-based campaign group, published findings on the extent of CCTV usage in schools.[3] They issued Freedom of Information (FOI) requests to 4,092 secondary schools requesting details on the total number of cameras on the premises, the number of internal cameras, and the number of cameras inside bathrooms and changing rooms. The FOIs elicited responses from 2,107 schools, 90 per cent of which had CCTV, with an overall average ratio of one camera to every 38 pupils. Some schools had much higher levels of coverage with a ratio of one camera for every five pupils. Extrapolating the findings Big Brother Watch estimated that there were over 100,000 CCTV cameras operating in Secondary schools and Academies in England, Scotland and Wales.[4]

School CCTV systems are hugely varied. Some schools utilise a handful of strategically placed cameras in specific locations, whereas others have scores of cameras monitoring virtually every area including classrooms, corridors and sports facilities. In the US, as long ago as 1994, it was estimated that a quarter of public schools had CCTV, and that nearly all of these (94 per cent) had CCTV in the classrooms.[5] Although less common, some schools have installed fully operative cameras in sensitive locations such as pupil's toilets. In 2004 the Bog Standard Campaign (aimed at improving school toilets) found that CCTV cameras in student's toilets in British secondary schools were 'quite common' and in 2008, the Association of Teachers and Lecturers (ATL) estimated 7 per cent of schools in England and Wales had CCTV in the pupil's toilets. In 2012, Big Brother Watch found that nearly 10 per cent of 2,107 schools responding to their FOI request had CCTV installed in 'changing rooms and bathrooms'.[6] Some schools have generated alarm (as well as police

DOI: 10.1057/9781137308863

intervention) by inadvertently filming schoolchildren changing clothing for sports activities.[7] There are schools using covert cameras, and those that wire their systems directly to the police station.[8] Many schools have equipped their CCTV cameras with microphones enabling audio monitoring to accompany visual footage. For example, a number of schools in Greater Manchester, England installed microphones alongside surveillance cameras and required teachers to wear earpieces so that observers could provide live feedback on their teaching delivery and performance.[9] Aside from potentially stifling the natural teaching experience, there are clearly legal and ethical issues raised.

Despite a pervasive presumption that CCTV is solely implemented to prevent crime it has found many non-criminal applications such as the invigilation of exams,[10] monitoring teaching performance,[11] curtailing general 'horseplay' and tackling the 'misuse of paper towels and soap' in pupils' toilets.[12] A British survey conducted by the ATL found that although 98 per cent of teachers claimed CCTV was installed for security purposes, 50 per cent reported it being used in other ways such as to monitor pupil behaviour.[13]

CCTV in schools has been esteemed by some as an impartial observer or a 'guardian of truth' in a growing litigation age that teachers can draw upon to verify events and elicit the truth. Simply the threat of accessing CCTV footage can operate as a 'truth serum' and result in students admitting misdemeanours (see Case Study Box below).

CASE STUDY: IF THE SHOE FITS ... CCTV AS A 'TRUTH SERUM'[14]

A primary school in England used its CCTV system, installed in every classroom, the canteen and reception area, to solve the case of a pair of missing shoes. The school bursar explained how CCTV was pivotal in the investigation:

'Someone hid a child's shoes and we found it on the tape so we knew where they were and who had hidden them. No one was owning up so we rewound the tape. If the teachers say, "I will rewind the tape" it makes them own up'.

In this way CCTV can be used as a truth serum to elicit honest responses from those under interrogation.

The girl who took the shoes was sent a warning 'yellow letter' from the school. Children who get three of these letters are suspended.

DOI: 10.1057/9781137308863

The mother of the 'guilty' girl thought that this use of the surveillance cameras was a disproportionate response to a childish prank.

'She took the shoes and hid them in a locker. We all do silly things and play jokes when we are children. They should have given her a chance – she's only eight.'

The case was also brought to the attention of the Information Commissioner's Office (ICO) which condemned the response as 'deeply intrusive' and raising 'privacy concerns for teachers, students and their parents'. Furthermore, in a statement the Office added:

'One person's prank is another person's distressing incident. But constant CCTV monitoring of all children in a class cannot be justified with reference to the need to address classroom disruption.'

This case highlights the intrinsic function creep that is set in motion once surveillance mechanisms are installed. The temptation to apply them to a forever widening scope of phenomena often proves too hard to resist and further embeds surveillance as the default response to discipline and control. In addition, the case highlights the inability of the ICO and the Data Protection Act 1998 that it is tasked to enforce, to regulate the inappropriate usage of CCTV and other surveillance equipment.

Pupils have collectively mobilized themselves in protest against the installation of CCTV cameras in their classrooms. For example, students at a school in Waltham Forest, United Kingdom walked out of their school following the introduction of CCTV because they felt it threatened their civil liberties. The pupils refused to return until they had received assurances that it had been turned off. Furthermore, upon their return to the school, the pupils wore masks to continue their protest. In other cases individuals have refused to return to the school until CCTV cameras have been removed from pupils' toilets.

Inside the school, children have resisted CCTV in three main ways: avoidance of areas monitored by CCTV cameras; restricting the ability of the CCTV cameras to identify them; and thirdly, by repositioning the cameras so they were no longer monitoring their behaviour.[15] However, avoidance of, or interference with the cameras has been taken to be harmless and humorous subversions of the operation of the CCTV within the school. Rather than a product of malice or criminal intention, it has been considered as a playful expression of rebellion that is common amongst

DOI: 10.1057/9781137308863

young people and can be likened to relatively trivial activities performed by a boisterous minority in schools such as scrawling a name on a desk or experimenting with the boundaries of the school uniform. Resistance is explored in greater detail in Chapter 3.

Despite the proliferation of CCTV, whether in schools, shopping malls or the cityscape, there is no concrete evidence to suggest that it is the panacea to the myriad problems it has been introduced to tackle.[16] CCTV evaluations conducted hitherto have provided mixed and often contradictory findings but the expansion of CCTV has continued unperturbed. In terms of schools, there has not been any reliable independent research conducted to assess whether CCTV can or does reduce incidents of crime, bullying or any of the other objectives it is presumed to assist with. In a sense it doesn't matter whether CCTV is effective or not since its appeal lies in the commonsense perception that it must do *something*, and at times of heightened anxiety surrounding school safety, something is markedly better than nothing.

Automatic Fingerprint Identification Systems

Whereas once one would perhaps only talk of finger-painting in the context of the school, it is now fingerprinting that is becoming increasingly common. The first reported use of an Automated Fingerprint Identification Systems (AFIS) in a school was in Minnesota, United States in 1999. Just over a decade later, it has materialised in schools all over the world, primarily used for purposes such as registration, library book borrowing and cashless catering. AFIS is becoming more common in many European countries such as Belgium and Sweden. However, some countries have taken steps to outlaw its use in schools, such as China and Hong Kong amid privacy concerns.

AFIS were first used in British schools in 2001. In 2002, it was claimed that approximately 1,000 schools had purchased the equipment, resulting in the fingerprinting of 300,000 children.[17] A web poll by the children's rights campaign group Leave Them Kids Alone (LTKA) estimated that more than 3,500 schools had bought fingerprinting apparatus from two Department for Education and Skills (DfES) approved suppliers: Vericool and CRB Solutions. Indeed the DfES and the Office of the Information Commissioner (the UK's independent authority responsible for the 'data privacy' of individuals[18]) were actively encouraging schools to fingerprint pupils as 'best practice' in the handling of information and personal data. Most recently, a national newspaper submitted FOI requests to 27 local councils asking whether the schools within their area had implemented AFIS. The responses illustrated that 111 secondary and 84 primary schools were utilising the technology. Extrapolating these findings and applying them across the 150 authorities, it was concluded that at least a third of all British schools are using AFIS.[19]

The Protection of Freedoms Act 2012 will ban all English schools from collecting pupils' biometric data without the consent of parents from September 2013. However, since the ruling will not be applied retrospectively, vendors are clamouring to have their products installed, ultimately increasing the level of AFIS in English schools. Despite the complexities involved in generating accurate calculations regarding the number of schools using fingerprinting technology, it is clear that it is now commonplace.

There are many ways in which AFIS is deployed in schools. It is often used to replace library cards when borrowing books from the school library and to automate the registration process. Pupils are required to place their enrolled finger on the digital fingertip scanner to verify identity and entitlement to services. Some providers claim that the software speeds up processes, although research has shown systems are often slow to recognise enrolled users. For example, it was reported at one school that the 'plexiglass of the scanner often gets scratched or broken when students, either inadvertently or in an act of sabotage, drop keys or other objects on to it.'[20] Sometimes the system stops recognising a pupil's print entirely and they are required to reregister at the school office, often missing lesson time to do so.

AFIS has also been implemented in schools to automate the process of purchasing food at lunchtime. Parents deposit money into pupils' catering accounts which are then debited by a child's biometric fingerscan at the point of sale. As with registration uses, enhanced speed of operations is claimed to be a primary objective. However, similar to above, some schools have disabled the system after it failed to recognise pupils' fingerprints, actually slowing down the process of purchasing food. AFIS has also been attributed with the ability to tackle bullying as it removes the stigma attached to pupils who are eligible for free school meals since there is no means of differentiating them, and it reduces the ability to bully a child for their lunch money. However, there are clearly numerous solutions to this conundrum that do not involve using the biometrics of

DOI: 10.1057/9781137308863

pupils, and coercion is often displaced to other methods such as forcing another pupil to make purchases on their behalf.

Despite claims that fingerprinting technology is the panacea to the broad range of problems that have apparently befallen pupils (obesity, bullying, truancy, poor timekeeping and so on), there has been no independent or verifiable research to support or negate the claims and the area has received relatively little attention by academic researchers. The lack of understanding about how these systems work, how secure they are and whether they are interoperable has enabled the vendors' discourse to become the dominant authority on the safety and utility of AFIS. Kim Cameron, a computer scientist and currently Microsoft's Chief Architect of Access, captures the way in which discourses on new technologies are often at the mercy of the sales pitch rather than the facts, when he writes:

> It drives me nuts that people can just open their mouths and say anything they want about biometrics... without any regard for the facts. There should really be fines for this type of thing – rather like we have for people who pretend they're a brain surgeon and then cut peoples' heads open.[21]

Independent empirical studies are desperately needed to fully understand the ability of these systems to fulfil their multiple objectives, as well as to understand the impact upon young people's perception of privacy, trust and responsibility.

Cashless catering and the ministry of food surveillance

AFIS and other cashless catering systems provide parents with control over what their children are able to purchase at lunchtime, and some vendors claim that this can tackle childhood obesity. Some systems are programmed to 'flag' when pupils attempt to buy items that their parents have specified as being unsuitable due to special diets, allergies or preference. If a pupil attempts to purchase contraband items an alert is triggered and a lunchtime organiser may intervene. Systems such as *My Nutrikids* enable parents to download weekly dietary reports detailing all their child's purchases, whereas some schools send the report home in the mail. In order to generate interest and compliance, some schools have introduced a points system for pupils who eschew unhealthy snacks which are converted into prizes at the end of the year.

It is not just 'cashless catering' that has placed eating habits under increasing surveillance; initiatives such as the 'the packed lunch toolkit'

DOI: 10.1057/9781137308863

devised in Gloucestershire, England involved pupils' lunchboxes being emptied out, photographed as 'evidence' and then ranked according to nutritional value.[22] Whereas Jamie Oliver's Orwellian sounding Ministry of Food caused resistance amongst parents who rejected what they perceived as militant food monitoring. They responded by feeding their children deep fried fish and chips through the school gates.[23] In a number of US states, selected students classified as overweight have been given a wristwatch-like device to count heartbeats, detect motion and track students' sleeping habits with the aim of reducing their weight.[24] The information is transmitted to a website accessible to the student and school staff. Even bodily functions are sought to be regulated with only specified times allowed for visits to the bathroom.

Radio frequency identification (RFID) systems

Radio frequency identification, or 'microchipping', is becoming common in schools around the globe. RFID tags, either passive or active, are embedded with a microchip containing electronically stored information with a unique identification number. 'Passive' tags do not contain their own power source such as a battery and therefore are much smaller with a virtually unlimited life span. They remain dormant until stimulated by a radio signal from an external reader device. Conversely, 'active' tags use a local power source to transmit and broadcast their data. It is claimed that the current technology provides passive tags with a reading range of '100 feet or more' depending on the frequency and power of the system. Active tags have a far longer range. RFID data is retrieved by readers using electromagnetic energy in a particular frequency. In both passive and active systems, the tag does not need to be within line of sight of the reader, 'the request and response process is both silent and invisible'.[25]

RFID has been trialled in schools in places such as the UK, the Philippines and South America, but it is Japan and some states in the US that are leading the way in the RFID tracking of pupils. RFID microchips are commonly embedded in school passes, which are required to be carried or worn at all times by pupils, whereas some schools embed the chips into 'smart uniforms' or use conductive 'smart threads' embroidered onto school jumpers.[26] As with CCTV, the objective of RFID is usually presumed to be to safeguard pupils, but a cursory look at how it is actually used belies this. In the Philippines, schools

DOI: 10.1057/9781137308863

use RFID tags embedded in school passes for borrowing books from the library, purchasing items, to monitor attendance and track pupils around campus. In Brazil, 20,000 grade-school students are wearing so-called 'intelligent uniforms' embedded with locator chips under the school crest. The education secretary claimed that the RFID was used to prove truancy levels: 'We noticed that many parents would bring their children to school but would not see if they actually entered the building because they always left in a hurry to get to work on time. They would always be surprised when told the number of times their children skipped class.'[27] Similarly in the US a key objective of RFID is to tackle truancy, but it is not motivated by a desire to ensure the safety of the child, or out of concern for educational attainment. The impetus is money. Schools are given their funding based on number of pupils attending school. Realising that pupils just needed to be on campus (rather than in class) to 'count', the RFID system provides a way of claiming the revenue even if students are not present for roll-call. For example, the Northside Independent School District expects to collect upwards of $2 million in state funding by improving 'attendance rates'. It is claimed that the RFID system cost around one-quarter of that sum to initiate and another $136,000 annually in maintenance, so it becomes an economically sound investment.

The RFID systems have generated concerns from civil liberties groups on privacy grounds as well as concerns that information transmissions can easily be 'skimmed' and 'eavesdropped' potentially placing vulnerable young people at risk. The American Civil Liberties Union's public education director stated:

> The technology was originally designed for shipping goods and cattle, not taking roll at school, thus RFID chips make the perfect stalking device. Because the technology is easy to acquire, it is vulnerable to hacking which could allow someone outside the school to monitor a student's off-campus whereabouts if they obtained the student's tracking number.[28]

In August 2012, privacy advocacy groups published a paper in which they argued that RFID tracking in schools violates students' rights to free speech and association because the technology tracks not only an individual's location, but it can monitor people who congregate together. In addition, the group claim that RFID tracking of pupils is dehumanizing, poses potential health risks and is serving to condition young people to intense monitoring of their behaviour and whereabouts.[29] Perhaps the

DOI: 10.1057/9781137308863

strongest objection to RFID in schools has come from religious groups (see Case Study Box in Chapter 3, page 72).

Policing the Surveillance School

Police officers have historically delivered educational and informational inputs, usually with a focus on road safety, the perils of drug usage and security awareness. The intention has typically been for police officers to augment learning and provide a law enforcement perspective to aspects of the curriculum that were perceived as having synergy with their role. However, what we see now in many urban schools is not just occasional police visits to schools, but rather the policing *of* schools. Increasingly police officers are present full time on campus, signalling what Jonathan Simon has identified as a merging of schools with the penal system.[30]

School police officers have their origins in Michigan, US and can be traced back to as early as the 1950s. The rampant take up of the approach and resultant nationwide implementation didn't emerge in earnest until the 1990s.[31] Police in schools has since become a common phenomenon in many countries, although the approaches vary with different duties, objectives and responsibilities.

In the United States, School Resource Officers (SROs) are the most notable segment of law enforcement in schools, and are usually uniformed and fully armed. The National Association of School Resource Officers (NASRO) was founded in 1991. The United States Department of Justice's Community Oriented Policing Services (COPS) has impelled the development by providing federal level funding for its 'COPS in Schools' (CIS) Program, designed to help law enforcement agencies hire SROs to engage in community policing in and around primary and secondary schools. COPS has awarded in excess of $750 million to more than 3,000 grantees to hire more than 6,500 SROs through the CIS program, and in addition provided more than $10 million to hire approximately 100 SROs through the Safe Schools/Healthy Students program, making 'school-based policing is the fastest growing area of law enforcement'. In 1997 there were just under 10,000 SROs assigned to public schools in the US, and a 1996–97 survey demonstrated that 22 per cent of schools had a police officer or other law enforcement representative stationed on the premises at some time during the school day.[32] To put this in context, in New York City in 1996, 3200 uniformed school security officers worked

DOI: 10.1057/9781137308863

in the Division of School Safety of the City Board of Education representing 'a contingent larger than the Boston Police Department'.[33] Clearly these numbers will have increased substantially since then. Rather than assessing the need for police presence in schools on a case-by-case basis they have come to be viewed as a necessary component of the new school discipline. Indeed, NASRO assert that 'the presence of local police on campus is essential.'

Some schools voluntarily relinquish discipline and rule enforcement to police and security officers, whereas in some locales the transferral is actually enshrined in law. For example, the NYC Board of Education, alongside many US states, have voted to transfer responsibility for school safety to the city police. Augmenting the blurring between school discipline and criminal justice, Arizona state law requires that school officials report any crimes or security threats involving students directly to the police.[34] Some US states have introduced legislation requiring schools to provide police with information about students, including personal data gathered by school therapists and councillors.

Police officers take on many forms: from uniformed armed officers patrolling the campus, to plain-clothed officers whose primary function is education and mentoring. There is also variance in how frequently and for what duration police officers are on campus. Some take in the school as part of their regular beat, whereas others are stationed full time on site and have their own offices. As with the take up of many surveillance practices, it is difficult to identify a typical approach or strategy. Complicating matters further, many schools use more than one type of law enforcement officer concurrently. For example, full time campus police officers might utilise support from local police or private security guards.[35]

The objectives of police on campus are varied, but they can be categorised into four main aims and objectives; clearly there may be overlap between them:

- *Crime control.* This can involve prevention measures such as providing target hardening advice, detection, responding to offences, as well as sourcing information about individuals and activities occurring off campus.
- *Discipline.* Providing input into discipline matters of a non-criminal nature.
- *Mentoring and building relationships.* A commonly cited aim of school police is to forge better relationships between youth and police, and eradicate negative perceptions of law enforcement.

DOI: 10.1057/9781137308863

- *Education.* A more traditional role of police in schools is to provide educational input on a range of issues such as drugs, personal safety, the law and road safety.

Police are typically engaged in a number duties and activities that fall within these four categories. In a study of SROs in four US public high schools, they were found to perform the following activities: observing and patrolling the corridors, completing administrative police work, investigating minor criminal incidents, mentoring students and participating and intervening in school discipline matters of a non-criminal nature.[36] In addition, police officers patrol school grounds, chaperone events and counsel students[37] and they do of course continue to provide teaching on law enforcement and related matters. Ericson and Haggerty noted in their study that classroom teaching formed the bulk of a police officer's school role, with the rest of the time spent performing as the security officer for the school which entailed dealing with an array of issues including disruptive children, trespassers and aggressive parents.[38] Some school police go so far as conducting 'lock down drills' and 'SWAT Team' rehearsals, perhaps believing that they are performing 'reassurance policing' to a concerned public.[39] Clearly this type of activity can be terrifying for young children as illustrated by the woefully misjudged lockdown and drugs raid at Stratford High School in South Carolina in 2003.

Whilst NASRO asserts that 'statistically speaking, the effectiveness of School Resource Officers is firmly established', there are mixed findings regarding the efficacy of police in schools, and a lack of an evidence base to draw upon. There is no doubt that schools should be crime free, but there are questionable findings as to whether onsite police is an effective means of fulfilling this. Many see the use of police on campus as a '"quick-fix" strategy' that, similar to zero tolerance policies, fails to get to root the causes of misbehaviour at school.

The seemingly benevolent funding awarded to schools by the US Justice Department might actually be for their own gains as research has found that one of the main beneficiaries of police in schools initiatives is the police themselves, often through the creation of informants. Students are not only a source of criminal misdemeanour but also a hitherto untapped resource of information and knowledge about illegal activities in the community.[40] In Canada, the Falconer report made more than 100 recommendations, one of which was the creation of a website on which

DOI: 10.1057/9781137308863

students could file anonymous reports of violence. The cultivation of students as informants has important implications for relations of trust, not only within the school, but also in the broader community.

Some studies have claimed that students are positive about the police presence in their schools, but recognise that there is a normative discourse in operation in which the policing of schools is becoming routine.[41] Illustrating this point, Kupchik and Bracy highlight a 'disturbing' report produced by the New York Civil Liberties Union documenting a range of misdemeanours such as the 'physical abuse and sexual harassment of students, retaliatory arrests of school staff who protect students from abuse at the hands of police, and other offences'.[42] Kupchik and Bracy point out that in this scathing report on the conduct, attitudes and behaviour of SROs in the US, there is no critical discussion about whether it is appropriate and desirable to have in-school police officers in the first place. Their presence has become a *fait accompli* that is not challenged even by the very organisations that lambast their conduct.

The infiltration of police personnel into the school environment forges a direct link between students and the criminal justice system. Behaviours that once would have been dealt with internally are passed on to the police and are more likely to result in arrest, and minor transgressions are increasingly criminalised. The 'constant link between schools and law enforcement officials lowers the threshold for entry to the justice system.'[43] Instead of internal discipline and policies for infraction, violations of the law are now much more likely to be extradited to criminal justice agencies external to the school, resulting in increased arrest and incarceration of young people. For example, in Texas, the police gave close to 300,000 'Class C misdemeanour' tickets to children as young as six in 2010 resulting in fines, community service and even prison time. What was once handled with a telling-off by the teacher or a call to parents can now result in arrest and a record that may cost a young person a place in college or a job years later.[44] Whilst it is argued that school-based police 'provides a vital bridge for potential young offenders'[45] to steer them away from a lifetime of crime, perhaps the only bridge facilitated by having onsite law enforcement is one that links the school directly to the prison. And it's not just police in the corridors; criminal justice personnel such as probation officers, and private security professionals have also been introduced further criminalizing schools and those that reside within.

Profiling

Children are increasingly filtered or 'sorted'[46] with every aspect of their existence and conduct monitored and controlled. Illustrating how social sorting is driven by fear, if not moral panic, in 1999 the US Federal Bureau of Alcohol, Tobacco and Firearms in conjunction with a 'threat assessment company' developed a pilot scheme called 'Mosaic'.[47] The programme was used to vet and rate potentially violent students based on a crude assessment of questions drawn from their case histories. Similarly, onsite police have been found to compile files on those students that they thought were potentially 'at risk'.[48] Following the Columbine massacre, the Federal Bureau of Investigation (FBI) released its own psychological examination entitled 'The school shooter: a threat assessment perspective'. Such profiling often draws upon simplified notions of perpetrators in which any signs of disenfranchisement, non-conformity and oppositional behaviour are construed as symptoms of a dangerous and violent mind. They often simply equate difference with dangerousness. Although the Secret Service, the FBI and researchers in the area of threat assessment have consistently found that it is impossible to construct reliable profiles that can be of assistance in promoting school safety, 'school-shooter' profiles continue to be devised and circulated.[49]

Zero tolerance approaches to school discipline

The introduction of criminal justice personnel to school campuses have brought with them new narratives of control that shape the way that schools are governed and pupils are managed. Zero tolerance approaches to discipline provide a case in point. Growing out of the police enforcement of drug trafficking in the United States during the mid- to late-1980s, zero tolerance has come to describe disciplinary philosophies and policies that are intended to deter disruptive behaviours through the application of severe, predetermined and certain punishment. In the school context it has been defined by the Zero Tolerance Task Force as:

> [A] philosophy or policy that mandates the application of predetermined consequences, most often severe and punitive in nature, that are intended to be applied regardless of the apparent severity of the behavior, mitigating circumstances, or situational context. Such an approach is intended to deter future transgressions by sending a message that no form of a given unacceptable behavior will be tolerated under any circumstances.[50]

DOI: 10.1057/9781137308863

Motivated by a belief in a growing culture of violence and indiscipline in schools, since the early 1990s, educators have adopted zero tolerance as a no-nonsense response to drugs, weapons and gang activity on campus. By 1993 incarnations of zero tolerance were being adopted by school boards across America and were cemented as national policy in US schools when it became enshrined in the Gun-Free Schools Act of 1994.[51] By October 1995 all US states were in compliance with the Act.[52] A national report examining the school year 1996–1997 found that 94 per cent of all schools had zero tolerance policies for weapons or firearms, 87 per cent for alcohol and 79 per cent used mandatory suspensions or expulsions for violence or tobacco.[53]

In Britain, whilst not enshrined in law or official policy, in 2005 the zero tolerance approach was catapulted onto the agenda when the then Secretary of State, Ruth Kelly, proclaimed:

> We must now have zero tolerance of bad behaviour in the classroom and create a culture of respect, of good behaviour and firm discipline, and this must be the norm in all schools in every classroom all of the time.[54]

The zero tolerance rhetoric has continued in Britain with the Education Secretary introducing the Education Act 2011 with: 'Our bill will put heads and teachers back in control, giving them a range of tough new powers to deal with bullies and the most disruptive pupils. Heads will be able to take a zero tolerance approach.'[55] The Act introduced a number of provisions increasing teachers' ability to search pupils without consent and confiscate possessions, including the seizing of electronic devices, viewing content stored in them and the ability to erase data or files from the device if they believe there is good reason to do so. Prior to the Act, they could search only for weapons, drugs, alcohol and stolen goods. Head teachers and authorised staff can use such force as is 'reasonable' when searching a pupil without consent for prohibited items. Perhaps as an indication of what is to come, the British government is studying the American experience in dealing with unruly young people and juvenile justice in the wake of the riots in England in August 2011. The justice minister visited Texas in September 2012 to study police programmes in schools, juvenile courts and prisons.[56]

The lack of a shared universal definition of zero tolerance renders it difficult to provide accurate figures regarding the extent to which this philosophy now underpins discipline in schools internationally. The approach is often inferred in the vernacular of the dominant disciplinary

DOI: 10.1057/9781137308863

style subscribed to, as with the UK political rhetoric above, even if it is not explicitly termed and classified as Zero tolerance.

Zero tolerance policies are promoted as an effective and just response to drug use or possession, carrying weapons on school premises and violence. The central tenet, and focus of controversy, is that students (and sometimes staff) that contravene the rules of the school are subjected to a mandated, usually severe, punishment. There is no room for discretion, mitigating factors or an appreciation of a variant scale of severity. Of course deliberate contravention of school rules, and more seriously the law, must be identified and dealt with in an appropriate manner but in order for discipline to be effective it must be fair and proportionate. Zero tolerance approaches dictate that minor and sometimes unintentional misbehaviour are treated in exactly the same way as serious and deliberate contraventions of the rules. Those in authority in the school are stripped of the ability to apply judgement or assess extenuating circumstances. The use of law enforcement principles in schools has resulted in significant net-widening and not surprisingly this has resulted in some imprudent and ludicrous outcomes. Mandated punishments have been applied to smoking, wearing the incorrect uniform and ambiguous categories of behaviour such as 'disruption'.[57] Pupils have found themselves at the receiving end of harsh punishments and even in trouble with the police for transgressions as innocuous as 'sharing aspirin', 'bringing nail clippers and scissors to class', donning a short skirt, possessing cigarettes, spraying perfume, inappropriate hairstyles and even being late for school.[58]

CASE STUDY: PEA SHOOTERS AND 'TERRORISTIC' BUBBLE MACHINES: ZERO TOLERANCE RESPONSES TO SCHOOL DISCIPLINE[59]

Bored and feeling slightly mischievous, 14-year-old Andrew Mikel used the casing of a ball point pen and the power of his lungs to launch small pieces of plastic across the room towards his fellow students at Spotsylvania High School, Virginia.

Andrew admitted it was a stupid thing to do but he did not expect his childish prank to impact the rest of his life. Unfortunately the school's disciplinary board interpreted his actions in a different way. His makeshift pea-shooter was reclassified as a Projectile Weapon and when the deputy sheriff was called to the scene his

DOI: 10.1057/9781137308863

misbehaviour was quickly escalated to a criminal offence. Mikel was charged with three counts of misdemeanour assault and was expelled from school. A civil liberties organization is appealing the case in Virginia state Court.

At the time of writing another overly zealous application of zero tolerance policies has made the news headlines in the United States. In January 2013, a school in Pennsylvania suspended a five-year-old girl for 'threatening' to use her *Hello Kitty* pink toy bubble gun (which was at home at the time of incident) to shoot bubbles at her classmates. The school described her actions as 'terroristic'. The girl was initially given a 10-day suspension which was later reduced to two days. The case is progressing towards the courts as her parents are considering suing the school in an attempt to have the incident removed from their daughter's permanent school record.

The federal Gun-Free Schools Act mandates that schools expel students who take weapons, including hand guns, explosive devices and projectile weapons, to school. In the wake of the tragic Sandy Hook Elementary School shooting in December 2012, schools in the United States have been particularly assiduous and severe in applying the principles of the Act to criminalize a broadening range of childish activities and behaviours.

As with many initiatives, zero tolerance approaches to school discipline are not predicated on efficacious evidence. Research on zero tolerance in the US found, somewhat counter-intuitively, that rather than providing a safer and more orderly school environment through the removal of unruly pupils, 'schools with higher rates of school suspension and expulsion appear to have *less* satisfactory ratings of school climate', and 'less satisfactory school governance'.[60] In contrast to the view that zero tolerance prevents further disobedience through expedited punishment, school suspension correlates with future rates of misbehaviour from those students who are suspended. The deleterious impact of suspension and expulsion on students is well documented. A 2011 study of school discipline in Texas found that students suspended or expelled for 'discretionary offenses' (those for which state law does not automatically call for an automatic suspension or expulsion) were twice as likely to repeat a grade as those who had not received such punishment. The study compared students from similar demographic groups and schools in

DOI: 10.1057/9781137308863

an attempt to isolate the effect of school discipline as much as possible, although it could not prove that time away from school directly caused the children to be held back.

Zero tolerance has been related to student shame, alienation, rejection and breaking of healthy adult bonds.[61] The disproportionate punishments meted out for minor infractions and indiscipline leave children feeling that the school system and its discipline regime is unfair, in all likelihood leading to withdrawal and marginalisation. When punishments are drastically out of kilter with the offence they lose legitimacy, thus they are likely, somewhat paradoxically, to increase student misbehaviour rather than reduce it. Drawing on the available evidence, Aaron Kupchik argues that 'one of the best ways to reduce school crime is to enhance students' perceptions that rules and punishments are just'.[62] Furthermore, blanket severe punishments can escalate the seriousness of an offence since the student knows they will receive the same punishment irrespective of how serious the act they committed was.

There are now calls to move away from zero tolerance to a more measured and context specific approach that takes on board the nuanced nature of rule contraventions, extenuating circumstances and the need for judgement and discretion. For example, in the US, Indiana, Texas and Virginia, have proposed legislation to modify zero tolerance procedures or expand the range of disciplinary options available to schools. Similarly, some councils in England are seeking a ban on short-term suspension after over-zealous usage premised of zero tolerance resulted in inflated levels of suspension and expulsion. Some education scholars favour an approach predicated on Restorative Justice to engage students and stakeholders in an interactive process to repair harm by addressing the causation of misbehaviour and consequential damage.[63] Although the rhetoric of zero tolerance may have subsided in recent years, the principles continue to influence policy and practice.

Screening and searches

Metal detectors

Globally, many schools require that pupils pass through security arches upon entering the school premises or use hand-held metal detector wands to perform random searches. Back in 1996–97 just 4 per cent of US schools conducted random metal detector checks and just 1 per cent

DOI: 10.1057/9781137308863

routinely screened students with a metal detector.[64] But since then a 'cultural readiness'[65] to implement checkpoint security in schools following a number of high profile violent incidents and changes in legislation have seen the use of metal detectors in schools become commonplace.

The US Safe Schools Act 1994 mandated that schools in receipt of federal funding must allocate it towards preapproved activities and purchases, one of which included '[a]cquiring and installing metal detectors and hiring security personnel'. In the US, metal detectors are most common in urban school districts. Approximately 12 per cent of high schools, 9 per cent of middle schools and 2 per cent of primary schools performed random checks with metal detectors in 2009–10.[66] Waltham Forest Council was the first in England to introduce a borough-wide weapons screening programme in schools, with knife arches in every secondary school.[67]

There are clearly logistical difficulties involved with screening all pupils. A report by the US Department of Justice found that in order to avoid huge waiting lines, schools had to completely restructure the school day, staggering class start times over a 90-minute period.[68] In addition, whilst the cost of metal detection equipment has reduced, the initial purchase is insignificant compared with the ongoing personnel costs required for operation. For example, in 1999 it was estimated that for one New York school with 2000 students, the weapons screening required nine additional security officers for approximately two hours each morning, equating to 100 security-officer hours per week.[69] The programme, not surprisingly, represented a substantial financial outlay.

It is not clear whether metal detectors are an effective means of preventing school violence involving weapons. The shooting of a 14-year-old student in early 2013 at the Price Middle School in Atlanta that was fitted with metal detectors raised questions about whether reliance on these systems to curtail weapons offence is the appropriate response. Since there is substantial evidence that young people carry a knife for protection,[70] the introduction of metal detectors may be counterproductive. Pupils may interpret the widespread introduction of metal detectors as evidence that their peers carry weapons, and so there is the very real possibility that they may decide to carry knives, or other weaponry, themselves. Furthermore, students are more likely to report feeling unsafe following the introduction of metal detectors and other 'jail-like security' measures at their school.[71]

DOI: 10.1057/9781137308863

Drug testing

In January 2005, assisted with funding from the *News of World* newspaper and support from the manufacturers Altrix Healthcare plc, a school in Kent become the first state school in the UK to report the introduction of random drug testing, opting for the use of saliva tests.[72] Drug testing is already widespread in independent boarding schools, with approximately three-quarters using some means of testing students for illicit substances. Beginning in the 2013–14 school year, students at Rockhurst High School in Kansas City will be mandated to undergo random drug testing. Students will be required to provide a staff member with 60 strands of hair which will be tested for a variety of substance use over the previous 90 days, including cocaine, PCP, opiates, methamphetamine, marijuana and binge alcohol.[73]

There are many examples of 'suspicionless' drug testing in schools around the world using different techniques including saliva tests, hair strand tests, sweat and urine samples. It has been argued that drug testing pupils can lead to a loss of trust in the school environment and lead to some pupils being negatively labelled. In some locations, pupils have resisted mandatory school drug tests by shaving off their body hair.[74]

It has been claimed that the cost of testing is likely to exceed most schools' allocated funding for drug education, prevention or counselling.[75] For example, in one school district in US, the positive detection of drug use for just 11 students costed $35,000.[76] Further demonstrating the exorbitant costs involved, the Australian National Centre for Education and Training on Addiction estimate that introducing nationwide drug testing of each child once a year in Australian schools would cost AUD$355 million for saliva tests or AUD$302 million for urine tests.[77]

Surveillance Schools are not limited to the above means of screening and searching pupils. Many now employ 'sniffer dogs' to search for drugs and weapons, whereas others have introduced transparent lockers and backpacks for students to assist with the scrutinising of their possessions. Pupils have been subjected to facial recognition, palm vein scanners and iris scanners. Student online activity, increasingly viewed as criminogenic, has become a focus of surveillance with internet tracking and a 'culture of over-blocking' websites.[78] There are many other types of surveillance taking place in schools, and too many to outline in detail here.

In presenting some of the features of the Surveillance School it is important to recognise that it is not just a case of more technologies

DOI: 10.1057/9781137308863

finding their way into schools, nor does it simply represent the automation of processes already in existence. Rather, the Surveillance School is characterised by a new culture of control that is facilitated and augmented by a constellation of surveillance practices that take on multiple forms: actions, behaviours, progress, beliefs and bodies are the objects of surveillance.

For every claim to effectiveness and enhanced student safety, there is a counter claim that the same outcome could be achieved for less money, with less impact on privacy and civil liberties and often with more efficacy. There is no independent or verifiable research that has explored the effectiveness of these technologies, but there are of course infinite accounts and anecdotes that 'prove' or 'disprove' their value within the educational system. It is less about the actual effectiveness, but providing a visual ritual against fear. Putting effectiveness aside, a key issue is how these technologies imbibed with ideology, discourse and rhetoric are transforming schools. What are the impacts that they are having on pupils, teachers and society? These are the real questions.

Notes

1 Robers, S., Zhang, J., Truman, J. and Snyder, T.D. (2012) *Indicators of School Crime and Safety: 2011*. National Center for Education Statistics. B.o.J. Statistics.
2 Taylor, E. (2010) 'Evaluating CCTV: why the findings are inconsistent, inconclusive and ultimately irrelevant', *Crime Prevention and Community Safety* 12: 209–32.
3 BigBrotherWatch (2012) *Class of 1984: The Extent of CCTV in Secondary Schools and Academies*.
4 Ibid.
5 US Department of Education (1994) *Fast Response Survey System, 'Survey on Advanced Telecommunications in U.S. Public Schools, K-12'*. National Center for Education Statistics.
6 BigBrotherWatch 2012.
7 Police seized video footage from Charlestown Primary School in Greater Manchester, England after parents protested about CCTV cameras filming children as they changed for gym lessons. Initially, it was staff at the school who had contacted the police to ask them to remove the protesting parents. However, having listened to the parent's allegations about inappropriate recordings, the police officers seized the school's CCTV footage and a computer hard drive. The school reported that the particular cameras causing

DOI: 10.1057/9781137308863

concern had been disconnected. 'School CCTV seized after kids filmed changing', *Manchester Evening News*, 6 November 2009.

8 Lewis, T. (2003) 'The surveillance economy of post-Columbine schools', *Review of Education, Pedagogy, and Cultural Studies* 25 (4): 335–55.

9 Qureshi, Y., 'Class CCTV comes under fire', *Manchester Evening News*, 11 March 2009.

10 'CCTV could be used in exams rooms', *BBC News*, 11 April 2008.

11 'Teachers watched on CCTV cameras', *BBC News*, 4 March 2009.

12 'School head defends toilets CCTV', *BBC News*, 27 January 2009.

13 ATL: *85% of teachers have CCTV in their schools and nearly 25% worry about hidden cameras.* Available at www.atl.org.uk/media-office/media-archive/cctv-in-schools.asp (accessed 04 September 2013).

14 Lewis, J. and Condron, S., 'Caught on a classroom spy camera – the primary school pupil who hid this girl's shoes', *Mail Online*, 11 January 2009.

15 Mccahill, M. and Finn, R. (2010) 'The social impact of surveillance in three UK schools: "angels", "devils" and "teen mums"', *Surveillance & Society* 7 (3/4): 273–89; Taylor, E. (2010) 'I spy with my little eye: the use of CCTV in schools and the impact on privacy', *Sociological Review* 58 (3): 381–405.

16 Taylor, E. (2010), 'Evaluating CCTV: Why the findings are inconsistent, inconclusive and ultimately irrelevant', *Crime Prevention and Community Safety* 12: 209–32.

17 Davies, S., "Law that favours disorder," *The Guardian*, 21 September 2002.

18 www.ico.org.uk (accessed 07 April 2013).

19 Clark, L., 'One in three secondary schools fingerprinting pupils as Big Brother regime sweeps education system', *Daily Mail Online*, 9 June 2010.

20 Indymedia (2006) 'Student fingerprinting – not so Vericool', Indymedia UK. Available at: www.indymedia.org.uk (accessed 19 September 2013).

21 Cameron, K. (2007) 'Just lie so you can sell your product', in *Kim Cameron's Identity Weblog*. Available at www.identityblog.com/?p=775 (accessed 19 September 2013).

22 '"Big Brother" school lunchbox scheme abandoned', *The Telegraph*, 2 July 2010.

23 'Parents feed pupils through gates', *BBC News*, 15 September 2006.

24 Linge, M., 'Schools "spy" on fat kids', *New York Post*, 15 January 2012.

25 CASPIAN, EPIC & the Privacy Rights Clearinghouse (2012) *Position paper on the use of RFID in schools.* Available at: http://www.spychips.com/school/RFIDSchoolPositionPaper.pdf (accessed 17 March 2013).

26 Milne, J., 'Pupils can be tracked round town with chips', *Times Education Supplement*, 11 May 2008.

27 Lehman, S., "Locator chips keep track of students in Brazil," *The Guardian*, 22 March 2012.

DOI: 10.1057/9781137308863

28 Johnson, C. (2013) 'Judge: Kids must wear locator chips at Texas school', *HLNtv.com*.
29 CASPIAN, EPIC & the Privacy Rights Clearinghouse Position paper on the use of RFID in schools (2012).
30 Simon, J. (2007) *Governing Through Crime: How the War on Crime Transformed American Democracy and Created a Culture of Fear*. Oxford: Oxford University Press.
31 Mayer, M.J., (2008) 'Fact Sheet #5: School Resource Officers (SROs)' Consortium to Prevent School Violence'. Available at: www.preventschoolviolence.org (accessed 08 April 2013).
32 *Annual Report on School Safety: 1998*. Department of Education and Department of Justice (United States).
33 Devine, J. (1996) *Maximum Security: The Culture of Violence in Inner-City Schools*, Chicago: University of Chicago Press. p. 76.
34 Beger, R.R. (2002) 'Expansion of police power in public schools and the vanishing rights of students', *Social Justice* 29: 119–30.
35 Casella, R. (2006) *Selling us the Fortress*. Oxon: Routledge.
36 Kupchik, A. (2010) *Homeroom Security: School Discipline in An Age of Fear*. New York: New York University Press.
37 Beger, R.R. (2002).
38 Ericson, V. and Haggerty, K.D. (1997) *Policing the Risk Society*. Oxford: Oxford University Press.
39 Hinds, L. and Grabosky, P. (2008) 'Responsibilisation revisited: from concept to attribution in crimecontrol', *Security Journal* 23 (2): 95–113, p. 99.
40 Ericson, V. and Haggerty, K.D. (1997). See also Kupchik, A. and Monahan, T. (2006) 'The New American School: preparation for post-industrial discipline', *British Journal of Sociology of Education* 27 (5): 617–31; Hope, A. (2009) 'CCTV, school surveillance and social control', *British Educational Research Journal* 35 (6): 891–907.
41 Casella, R. (2006); Kupchik, A. and Monahan, T. (2006).
42 Kupchik, A. (2010), p. 79.
43 Kupchik, A. and Monahan, T. (2006), p. 622.
44 Mcgreal, C., 'The US schools with their own police', *The Guardian*, 09 January 2012.
45 Chamberlin, T., 'Guns in Schools', *The Sunday Mail*, 30 September 2012.
46 Lyon, D. (2003) *Surveillance as Social Sorting*. London: Routledge.
47 Staples, W.G. (2000) *Everyday Surveillance: Vigilance and Visibility in Postmodern Life*. Oxford: Rowman and Littlefield Publishers.
48 Kupchik, A. (2010).
49 Lewis, T. (2003).
50 Skiba, R., Reynolds, C.L.R., Graham, S., Sheras, P., Conoley, J.C., and Garcia-Vazquez, E. (2006) 'Are zero tolerance policies effective in the schools?

DOI: 10.1057/9781137308863

An evidentiary review and recommendations', American Psychological Association Zero Tolerance Task Force, p. 26.

51 Ibid.

52 Yell, M.L. and Rotalszki, M.E. (2000) 'Searching for safe schools: legal issues in the prevention of school violence', *Journal of Emotional Behavioral Disorders* 8: 187–209.

53 *Violence and Discipline Problems in U.S. Public Schools: 1996–97.* Washington, DC: U.S: Department of Education, Office of Educational Research and Improvement, National Center for Education Statistics.

54 Cited in *Beyond Punishment: Reframing Behaviour in Schools.* CfBT Education Trust, 2009, p. 12.

55 'Teachers able to confiscate mobile phones to control disruptive pupils', *The Telegraph*, 03 July 2010.

56 Mcgreal, C. (2012).

57 Skiba, R. and Peterson, R. (1999) 'The dark side of zero tolerance: can punishment lead to safe schools?', *Phi Delta Kappan* 80 (5): 372–82.

58 Lozada (1998) cited in Beger, R.R. (2002).

59 Sieff, K., 'Plastic pellet incident at Va. school ends in expulsion, assault charges', *The Washington Post*, 1 February 2011; Demarche, E., 'Pennsylvania girl, 5, suspended for threatening to shoot girl with pink toy gun that blows soapy bubbles', *FoxNews*, 19 January 2013.

60 Skiba, R. et al. (2006).

61 Ibid.

62 Kupchik, A. (2010), p. 4.

63 Stinchcomb, J.B., Bazemore, G., and Riestenberg, N. (2006) 'Beyond zero tolerance: restoring justice in secondary schools', *Youth Violence and Juvenile Justice* 4 (2): 123–47.

64 Riley, R.W. and Reno, J. (1998).

65 Jones, R. (2009) 'Checkpoint security: gateways, airports and the architecture of security', in H.O.G. K. F. Aas, and H. M. Lomell (eds), *Technologies of InSecurity: The Surveillance of Everyday Life.* Oxon: Routledge-Cavendish, 81–99, p. 92.

66 Robers, S. et al. (2012).

67 'Council launches knife arches in school', *The Telegraph*, 29 April 2009.

68 Reno, J., Fisher, R.C., Robinson, L., Brennan, N. and Travis, J. (1999) *The Appropriate and Effective Use of Security Technologies in U.S. Schools.* Office of Justice Programs, National Institute of Justice: US Department of Justice.

69 Ibid.

70 Phillips, A. and Chamberlain, V. (2006) 'MORI Five-Year Report: an analysis of Youth Survey Data' *Youth Justice Board*, www.yjb.gov.uk; Wilson, D., Sharp, C., and Patterson, A., (2006) 'Young people and crime: findings from the

2005 offending, crime and justice survey', Home Office *Home Office Statistical Bulletin*.

71 Mayer, M.J. and Leone, P.E. (1999) 'A structural analysis of school violence and disruption: implications for creating safer schools', *Education & Treatment of Children* 22 (3): 333–56; Lewis, T. (2003), p. 338.

72 Gerada, C. (2005) 'Random drug testing in schools', *British Journal of General Practice* 55 (516): 499–501.

73 Foxnews.Com, "Missouri high school reportedly to use hair samples for random drug tests," 31 January 2013.

74 Gerada, C. (2005).

75 Ibid.

76 American Civil Liberties Union (2002) *ACLU drug testing cases across the nation. Factsheet 15*. Available at: http://www.aclu.org/DrugPolicy/DrugPolicy.cfm?ID=11001&c=79 (accessed 18 March 2013).

77 Roche, A.M., Pidd, K., Bywood, P., Duraisingam, V., Steenson, T., Freeman, T., and Nicholas, R. (2008) 'Drug testing in schools: evidence, impacts and alternatives', *Australian National Council on Drugs*.

78 Hope, A. (2008) 'Internet pollution discourses, exclusionary practices and the "culture of over-blocking" within UK schools', *Technology, Pedagogy and Education* 17 (2): 103–13.

DOI: 10.1057/9781137308863

2

Visions of Control: A Case Study on School CCTV

Abstract: *CCTV represents the most common surveillance technology currently utilised in schools. Introduced for crime control purposes, it has undergone immense function creep and is now focused on a multitude of behaviours and activities such as truancy, smoking, bullying, teacher performance and minor classroom naughtiness. Pupils feel that their privacy is being invaded and that they are being criminalised.*

Keywords: CCTV, privacy, trust, awareness, resistance

Taylor, Emmeline. *Surveillance Schools: Security, Discipline and Control in Contemporary Education.* Basingstoke: Palgrave Macmillan, 2013.
DOI: 10.1057/9781137308863.

DOI: 10.1057/9781137308863

> If I wanted to be on Big Brother I would have auditioned for it
>
> (Alyssa, City Comprehensive)

This chapter presents findings from empirical research conducted at three secondary schools located in the North of England in 2005. The study aimed to provide an in-depth micro-sociological account of how CCTV shapes everyday interactions and alters the environment of the school, moving the understanding of CCTV beyond its narrow conception as a crime control technology. The schools have been given pseudonyms to preserve anonymity: Single Sex Comprehensive (SSC), City Comprehensive (CC) and Urban High (UH), and all participants' names have been changed.

Single Sex Comprehensive

Single Sex Comprehensive is located three miles from a busy city centre. At the time of research it had approximately 1300 male pupils, the majority of which were drawn from areas suffering from high levels of poverty, deprivation and crime. Typically, over 90 percent of students are from ethnic minorities and more than half speak English as an additional language. A 12-foot-high perimeter fence flanked the campus, although in places this had been torn down. Visitor access was controlled via an electronic gate supported by CCTV and intercom. Manoeuvrability around the school was limited by an electronic fob system that operated the internal doors to a number of corridors.

The CCTV system comprised 33 cameras located in classrooms, the dining hall, corridors, locker areas and changing rooms, although notably there were no cameras in the pupils' toilets at the time of research. All external cameras were capable of 360-degree vision and featured pan, tilt and zoom (PTZ). Cameras recorded 24-hours a day and the system was supported by mobile phones and handheld radio transceivers operated by the site managers. The monitoring hub was located in the school reception area and featured a nine-split screen and one single screen for targeted observations. No one individual held responsibility for the operation of the CCTV, but it could be accessed from any computer in the networked system by any member of staff.

DOI: 10.1057/9781137308863

City Comprehensive

City Comprehensive had more than 1,000 pupils, a large number of which were eligible for free school meals indicating higher-than-average deprivation. There were numerous minority ethnic groups represented and the number of pupils with a statement of Special Educational Need (SEN) was above average. Similarly, the number of 'looked after children' was regarded as high by the most recent Ofsted report, as was the number of pupils who joined and left the school during the academic year.

Approximately 85 cameras, both internal and external, had been installed in two waves, consisting of what was described as 'old' and 'new' cameras. There had been much deliberation regarding the placement of the 'new' cameras to maximise coverage, as the design of the 'old' system was regarded as 'ill-conceived' and patchy. Cameras were located on corridors, in some classrooms (typically those with expensive apparatus and equipment), the canteen and in the pupils' toilet area, although it was stressed by a Site Manager that when cubicle doors were closed the cameras could not see under or over them. City Comprehensive was the only school in which a member of staff had explicit responsibility for managing the CCTV. The Site Manager described this role as taking up 'a large chunk' of his working day. There was a central monitoring hub housing the recording equipment, but images could also be accessed via several monitoring stations located around the school.

Urban High

Urban High had approximately 1500 pupils, with an additional 350 students attending the adjoining Sixth Form College. The pupil intake was predominantly middle class and the school was very popular with parents resulting in it being heavily over-subscribed each year. Although a range of ethnic backgrounds were represented, the majority were White British and command of English was 'very secure'. The number of pupils who had special educational needs was below the national average. Ofsted described it as a 'very good school' with some 'excellent features' and standards of attainment were above average.

The school building had recently been replaced with a state-of-the-art building that drew upon Crime Prevention Through Environmental Design (CPTED) principles, including 'natural surveillance'. Large

DOI: 10.1057/9781137308863

internal windows were used for senior teachers' offices allowing them to overlook the corridors in panoptic fashion. The school had approximately 60 cameras located in some classrooms, the canteen and on corridors. Similar to City Comprehensive, the use of CCTV extended to the pupil's toilet area. Cameras were managed from a central control centre but networked computers could be used to access footage by senior staff.

Awareness of CCTV and its focus

Understanding awareness of CCTV is important as it precurses many of its presumed attributes. It is inextricably linked to the potential for CCTV to be a deterrent against criminal activity or for it to impact on expectations of privacy. Evaluations have often found that CCTV does not have the intended outcomes because awareness is often temporary or transient. In order for CCTV to alter the social environment, users of the space must first be cognisant of the presence of surveillance cameras.

The consensus amongst teachers was that it was 'easy to forget' about the cameras. They became habituated to their presence which served to deplete scepticism about their use:

> I don't really take heed of them now. They're just there. You don't notice them, I don't worry about them... I was more aware of them when I first started [working here] but you just get used to them. I don't notice now, but now that I've spoken to you I'll probably get all paranoid again [laughs]. (Ms Parker, CC)

Conversely, pupils were far more aware of CCTV and many claimed to know the exact positioning of cameras. Following some deliberation, one group indicated that they could determine where the cameras were located in their school, and perhaps more importantly, where there was not any coverage:

> [Steven, UH] If I was walking around [the school] now I could say where the cameras were. I couldn't sit here and list all the places, well probably most of them actually... I'm saying you know when there is one coming up. If I'm walking down the corridor, I know the places where they are. I know where to look for them.
>
> [Simon, UH] Yeah, that's true. I know the places where there aren't any. Like I know that there aren't any on the stairs and I know that there aren't any where everyone goes and smokes.

DOI: 10.1057/9781137308863

Even if the pupils were unable to recall the exact positioning of the cameras they would check to see whether there was CCTV in the vicinity before misbehaving.

> Sometimes you might not know exactly where they all are but if you were gonna do something then you'd have a look about [...] What I'm saying is; you don't need to know exactly where they are cos you can just look up to check. (Ajay, SSC)

Despite some awareness that there were gaps in coverage as outlined above, some pupils perceived the camera surveillance to be near omnipresent.

> I couldn't tell you how many there are, but there's a lot! You're pretty much always gonna be on one of the cameras. It's a bit weird thinking about it now, like someone could be watching you whenever they want. [Milly, CC]

In all three schools the pupils identified themselves as the intended recipients of the cameras' gaze, which goes some way to explain their relatively high level of awareness. On the other hand teachers did not regard themselves as subjects of surveillance. Similarly in a British survey on school CCTV usage, just one percent of teachers believed that CCTV was being used for staff monitoring and none believed it was used for 'teacher training or observation'.[1] In all likelihood as the Surveillance School has evolved this perception will shift as teachers become increasingly ensnared in the surveillance apparatus, particularly as discipline is outsourced to external agencies such as the police.

Understanding CCTV: aims and uses

Crime control and safety

Teachers invariably cited variants of 'crime control' as the main objective of CCTV in schools, although understanding of how it was used in this way was often nebulous and broadly conceived. For some, CCTV was understood to be a deterrent that could prevent criminal acts, or 'at least make *them* think twice' (Ms Bennett, UH, emphasis added). Although crime control was considered to be the primary objective of CCTV there were very few examples offered other than vague references to 'break-ins' and 'vandalism'.

DOI: 10.1057/9781137308863

Single Sex Comprehensive had a more concrete example of how CCTV was being used on campus (although the technology raised frustrations rather than solutions). CCTV was drawn upon in a bid to alleviate the threat posed by intruders coming onto the school premises. Historically, the school had a problem with local gangs and the campus often became a site where contentions would be played out as Mr Baker, a teacher, explained:

> The risk from pupils within the school is minimal. The risk here is that someone from outside will come in. We get a lot of ex-pupils coming in, occasionally in cars which will fly down the drive, whiz round and then they are off again. You get older boys hanging around outside waiting for youngsters coming out. We presume that it is gang-related [...] Our staff put themselves on the line all too frequently. The CCTV helps us to manage the situation, monitor what is taking place and compile the evidence that we need.

Mr Baker continued to outline how the intruders 'couldn't care less' about the cameras and so any deterrent effect was negligible. The alleged gang members would deliberately attempt to hide their identity by using items of clothing such as bandanas, hats and hooded tops to avoid identification:

> They have scarves up to their eyes and their hoods up. You get to recognize them, the same characters, but you can't *identify* them if you understand what I mean? Even capturing them on the cameras is useless most of the time in respect of identifying them to the police.

Techniques of evading CCTV detection through the use of clothing such as baseball caps, scarves and hoods are well reported[2] and it has been conceded that criminals have little fear of the cameras due to such simple avoidance. Despite the limitations of CCTV, there remained a prevailing presumption amongst teachers that CCTV *must* enhance both perceived and actual levels of safety.

> It can be useful for managing the school: to ensure that the pupils are safe and in the right place. If something were to happen such as a fire, or some people were to get on the premises with the intention to cause harm then CCTV may be our first line of defence. (Mr Jacobs, SSC)

In stark contrast, the pupils reported that CCTV did not make them feel safer. They preferred the physical presence of a member of staff, particularly for serious events or those requiring urgent intervention:

> It should be actual teachers looking out for bad behaviour, not cameras. Cameras do not stop people from doing things they shouldn't. The cameras

> aren't watched [...] they just record most of the time. What good is that? It's not going to stop people [from] being naughty. If a teacher was watching through the camera then they are not close to the place they are watching so it'll take them a while to get there. The fight, or whatever, will probably be over by then. The damage will have been done. (Brendan, CC)
>
> If someone got stabbed in school, they would never get a teacher there in time to stop them. And anyway, if that kind of thing is going to happen then it will whether there are cameras there or not. It could happen anywhere: on the way home from school, at the bus stop, in the park. Those cameras don't do nothing. There are loads of fights still. (Milly, CC)

Furthering this point, pupils were confused as to how the cameras could make anyone feel safer mainly due to the fact that the cameras were not proactively monitored.

> [Ben, CC]: I don't think it makes you feel safer. Why would it? What's it gonna do? It might record you being beat up or having summat taxed [stolen] but it's not gonna stop it. It's not gonna give you your things back.
>
> [David, CC]: Maybe it could if it identifies who took things from you?
>
> [Ben, CC]: Yeah, maybe; but I'm just saying it won't stop it in the first place [...] I wouldn't report it. It could just make it worse.

Although CCTV could potentially prove the identity of an assailant, this was not always regarded as beneficial to the victim who might fear reprisal. In this vein, the teachers' fairly optimistic assessment of CCTV protecting the pupils is actually reliant on external processes beyond the cameras, i.e. the willingness of the victim (or others) to report the incident. When CCTV is used mainly as a reactive forensic tool, its deterrence capabilities are likely to be minimal as recognised by the students. Of course perceptions of safety and actual increases in safety are very different. As this study did not seek to identify changes in the actual frequency of incidents, it is impossible to comment on whether the perception amongst teachers (CCTV improved pupil safety) or pupils (CCTV did not increase their safety) was more accurate. Such mixed views are similarly reflected in evaluation research on the effectiveness of CCTV.[3]

CCTV and displacement

Spatial displacement is when a crime is shifted from one location to another, typically because of 'target hardening' (referring to the increased security mechanisms applied to vulnerable locations or items).[4] An individual may decide not to commit a criminal act in an area covered by CCTV, but rather than being fully deterred they simply move beyond the cameras' field of vision. Evidence of displacement has been found in most CCTV evaluations[5] although some researchers have attempted to refute these findings.[6] The displacement of deviant or illicit behaviour was identified by pupils in their discussion of CCTV, particularly in relation to smoking.

> [Becky, UH]...I know that there aren't any [cameras] where everyone goes and smokes.
>
> [Ryan, UH] That's true. Why don't they put them there – everyone knows that's where people go to smoke and to wag it [play truant], so surely they'd be better off there than in the middle of the corridor where no-one is gonna do anything anyway?
>
> [Becky, UH] But that's why people go to smoke there – cos they are not there. If you put them there, then they'd just go and smoke somewhere else.
>
> [Ryan, UH] I guess so, but then they'd be off the school premises, so in a way it wouldn't matter.

The pupils perceived deviant activities as location specific rather than intrinsically wrong, and of little relevance once they had moved beyond the school gates. The extract supports the assertion that CCTV displaces deviant activities, rather than preventing them in their entirety. However, it is important to note that evaluations of CCTV consider displacement over much greater areas than a school campus. For example, they look at the displacement of car crime from one car park to another located some distance away, or from one town centre to a neighbouring town centre.

Bringing young people and discipline into focus

It soon transpired that despite the main function of the CCTV system being cited as 'crime control', on a day-to-day basis it was mainly used to monitor and discipline the pupils.

> [Mr Coles, UH]: The main purpose, I suppose, is as a deterrent against people being naughty in school. Stopping fights and things like that in the corridors or outside. Stopping them getting up to no good.
>
> [ET]: Any people in particular?
>
> [Mr Coles, UH]: Yes, the pupils. I would hope the teachers are behaving!

Teachers were perceived to be exempt from the scrutiny of the cameras as they were considered to conform unquestionably to the accepted

DOI: 10.1057/9781137308863

behaviour codes of the school. Indeed, the very suggestion of their not doing so evoked a humoured response of mock outrage from the teacher. Furthermore the possibility of CCTV cameras being installed in areas over which teachers had a sense of ownership, the staff toilets or the staff room, evoked strong objection.

Pupils were also of the view that CCTV was used in their school to 'keep an eye' on them and to ensure 'that they aren't getting into trouble or breaking the law' (Rowan, SSC). It was clear that CCTV was being used as part of the disciplinary apparatus aimed at ensuring compliance and good behaviour from students, rather than to detect, prevent or control criminal activities. A British survey similarly found that CCTV was often used for 'pupil behaviour monitoring' and 'pupil behaviour control'.[7]

A pupil at Urban High commented about the use of CCTV in the pupils' toilets: 'you wouldn't get that in the teachers' toilets', and similarly at City Comprehensive:

> There are cameras in our toilets. I don't think there are any in the staff toilets. I can't imagine it anyway. They are there to stop smoking and fighting, or people hiding in there to skip lessons, so you don't need them in the teachers' toilets for that. (Milly, CC)

Not only were the cameras disproportionately targeted towards pupils in the school, but pupils were fully aware of this discrepancy. The targeting of some individuals over others based on the prejudice of operators has previously been reported[8] but the playing field has always been level in the first instance with the system providing an equal chance of surveying any individual in a given space regardless of their age, ethnicity, gender, dress and so on. In these studies, discrimination was purely the result of the camera operators' propensity to derive suspicion from stereotypical interpretations of an offender's appearance and behaviour. The emergence of CCTV in the school environment and its imbalanced application to inhabitants of that school, however, represents an intrinsic prejudice in the design and application of CCTV. Here we have a system that is inherently biased towards certain populations of the environment that it surveils, presenting interesting parallels to other institutions that are underpinned by control such as the prison in which inmates are the main focus of monitoring, and the hospital in which patients are the primary subjects of observations.

The ability of CCTV to exert an 'automatic functioning of power'[9] was inferred by some pupils.

DOI: 10.1057/9781137308863

> It encourages you to behave because you don't know if you are being watched through the camera or not. You're probably not, but you never know. I know people who just don't care now 'cos nothing has ever happened to them – there's no one watching. (David, CC)

Importantly, particularly for the theories examined in Chapter 4, the pupil indicates that the cameras do not impact equally upon all pupils. Some individuals have come to recognise that rarely are displays of nonconformity or delinquency in view of the cameras responded to, thus challenging the view that 'the surveillance is permanent in its effects, even if it is discontinuous in its action'.[10] The deterrent effect presumed by one never knowing whether they are being watched can actually work the other way round. Aware that CCTV cameras are not constantly monitored, or that the operator has numerous screens to observe at any given time, the rational deviant concludes that in all probability they are unlikely to be the focus of attention. Rather than individuals behaving in a manner deemed acceptable because they are never sure whether they are under surveillance or not, the deviant or criminal is prepared to take the risk. This actuarial approach was encapsulated in the comments of two pupils at Urban High:

> There isn't anyone sat watching the cameras all the time, and anyway if there was they would have to be looking at you, in your classroom at exactly the time that you were doing something wrong. It's a really big school so the chance of them catching something as it happens is really, really unlikely. I don't think the cameras work in that way. Maybe if there was one person watching one camera monitoring one classroom then they might work. Then a pupil might think 'hang on, I might get caught on that camera'. Even then there is what, 35 people to watch? [Tim, UH]
>
> Teachers miss things happening all the time and they're in the same room! [Becky, UH]

Teachers were also mindful that the supposed deterrent capability of CCTV was limited:

> The kids don't care [about the CCTV]. They know nothing's gonna happen. (Mr Jacobs, SSC)
>
> Some students might not play up because they know that the camera is on them. I guess it could improve behaviour, but then again I don't think they care. If they are going to be naughty I don't think a camera is going to stop them really. Even if it stops them in front of the camera then they will just move it elsewhere. (Mrs Williams, CC)

DOI: 10.1057/9781137308863

Mrs Williams went on to say that she thought it 'might stop something more serious like stealing equipment but it wouldn't stop general messing about'. Rather than having a blanket effect, the presence of the CCTV cameras only serves to deter some misbehaviour, and furthermore only those actions that can be visually identified such as stealing equipment.

Similar evidence that any disciplining power imbued in CCTV in schools is not permanent in its effects is provided by the quotes from Urban High pupils above regarding smoking. They outline how deviant behaviour takes place beyond the realm of the cameras thus inferring that the effects of CCTV may only be temporary. It has been found that the impact of CCTV in the city centre 'may be a strictly temporal and spatial phenomenon with those individuals with deviant intentions shifting the time and place of their activities to outside the cameras' gaze'.[11] It would appear that this is also true of the school.

The above examples challenge the view that visual surveillance in schools automatically results in compliance with the prevailing rules, which in turn raises questions for interpretations of surveillance focusing on the Panopticon (explored in Chapter 4). However, it should be borne in mind that the majority of pupils do generally conform to the rules of the school.

A guardian of truth in a litigation age

A recurrent theme to emerge was the view that CCTV could be used to protect teachers from false allegations by students, parents or other staff. The premise was that CCTV could present a value-neutral reflection of reality should a teacher's actions be called into question; 'if something happens we can check who was there and what was going on' (Mr Barry, CC). One teacher attributed CCTV in schools to a growing 'litigation culture':

> In our day and age the CCTV becomes a necessary evil. No one wants it, but within this culture of litigation and allegation it is necessary to protect oneself. It is necessary to document the true version of events. It cannot be one word against another, as it will be battled out in the courts and the truth doesn't always prevail. You have to be really careful and having a verifiable version of events has just become an unfortunate consequence of the society we now live in. I certainly would not want that taken away. (Mr Jacobs, SSC)

CCTV was viewed as a 'guardian of truth', an impartial medium that could be called upon should teachers need to negate false allegations

DOI: 10.1057/9781137308863

thrust upon them. In a similar vein, some teachers approved of CCTV in schools because they felt it was a useful resource to draw upon to force pupils to provide a truthful version of events.

> Sometimes you might mention the cameras if an incident has taken place and you are not yet sure of the facts. It can make people own up. They'll deny everything and then you say 'OK, I'll go and watch the CCTV footage then'. They'll go 'no, no, no' because they realise that they would be in a lot more trouble if they lied and then they were found out. It can be useful in those situations. You can definitely call their bluff. (Mr Coles, UH)

CCTV provided an 'all seeing eye' that could verify or disprove a pupil's version of events and simply referring to possible CCTV footage was enough to elicit confessions from some pupils, similar to the case study in Chapter 1.

The impact of CCTV in schools: privacy and trust

Privacy, trust and the use of CCTV were all inextricably linked for pupils; surveillance undermined privacy and was regarded to be a manifestation of mistrust. Pupils were incensed about the lack of trust that they were afforded by the school as well as by wider society. There were discernible differences of opinion between teachers and pupils. In general, teachers did not think CCTV presented a concern with respect to their privacy whereas the majority of pupils claimed that CCTV affected their privacy to some degree. As before, this finding is likely to stem from the perception that the CCTV cameras are there to monitor the pupils rather than the teachers.

Pupils described how the use of surveillance, in particular CCTV, in their school was symptomatic of an underlying mistrust of them:

> If you had a shop and you trusted a customer then you wouldn't watch them closely the whole time that they are in the shop. It happens to my dad just because he looks a certain way, he's got a beard and he wears baggy clothing. It's the same thing. If we were trusted then they wouldn't have so many cameras everywhere to watch us all of the time. (Mohammed, UH)

> If you want pupils to act responsibly then you need to show them that they are trusted. You need to treat them like adults, and with a little bit of respect. For some it becomes a self-fulfilling prophecy. If you are always expecting them to be up to no good then they might decide that they might as well misbehave because they are being treated like they are doing anyway. (Sarah, UH)

DOI: 10.1057/9781137308863

> People just presume that you are going to cause trouble. It's really unfair. I think, with the cameras, it's almost like saying that you have to be watched 24 hours a day otherwise you're going to misbehave. (Chloe, UH)

Mistrust was perceived to breed misbehaviour amongst some of the pupils in a classic exposition of Howard Becker's labelling theory whereby individuals come to act in accordance with the way in which they have been labelled or prejudged. The consequence of mistrust for Sarah culminates in the rational decision to engage in illicit behaviour because 'they are being treated like they are doing anyway'. The pupils believed that they were perceived as inherently deviant and required control to curtail their inevitable misbehaviour. This self-fulfilling prophecy can create a 'morality of low expectation'[12] in the school and result in a focus on managing pupils rather than engaging them in active learning and critical thought.

It can be inferred from the quotations above that surveillance is equated with mistrust since to demonstrate trust is not to surveil. Furthermore, it is subtly intimated that surveillance in the school is primarily focused upon the pupils because to be treated 'like adults' means that they do not disproportionately have their privacy invaded. Pupils perceive CCTV as an embodiment of suspicion; the very presence of the cameras is regarded as a symbol of mistrust and enough to galvanise a self-fulfilling prophecy.

Students identified the negative portrayal of young people in the media as one of the reasons they attracted disproportionate levels of surveillance.

> Every day the papers have stories of gangs, stabbings, muggings. It's just a small minority but it paints the picture that all young people are bad. There are a lot of old people that commit crimes. Imagine if the papers were painting the picture of all old people being bad – there'd be uproar. But because we are young then they can get away with it. (Cait, UH)

Pupils claimed that they were discriminated against because they had relatively less power. This is explicit in the perceived ability of adults to 'get away' with prejudicial attitudes and discriminatory behaviour against young people.

David, a pupil at City Comprehensive was quick to make the link between the media representations of youth and the increased use of CCTV in schools when he stated 'no wonder they got this place wired up like a prison – they think we're all crims'. References to the school being

DOI: 10.1057/9781137308863

like a 'prison', and there being a 'moral panic' about youth, demonstrates that some pupils perceived the increase of CCTV in schools was partly due to the demonisation and criminalisation of young people.

The pupils claimed that certain styles of dress had become shorthand for deviancy, particularly how young people wearing hoodies were susceptible to being labelled as deviant or criminal (further explored in Chapter 4).

> It's to do with how you look. If you dress the wrong way then you're gonna get hassled. If you're walking down the street with five or six of your mates and you're all dressed in hoodies then people are probably gonna think that you're about to mug an old lady. (Peter, UH)

As outlined above, there was a tangible perception amongst pupils that they were powerless to change the way in which they were perceived and the way that they were treated by the school, and by wider society. There was clear resentment that they were treated suspiciously or labelled as 'bad' simply because of their status as 'youth'.

There was resignation amongst pupils to the fact that within the school environment they had very little power to object or question the surveillance that they were subjected to. In this respect they fully internalised the authority of the school and accepted the apparatus used to observe and control pupil behaviour. Although there was awareness that surveillance was unequally administered and this was regarded as unfair, it was also expected and perceived as inevitable. The pupils resigned themselves to the fact that due to their status as 'young people' they were to accept the dominant control within that environment.

CCTV, toilets and privacy

Discussion of privacy mainly focused on the installation of CCTV cameras in the pupils' toilets at City Comprehensive and Urban High. It was stressed that cameras at both schools were positioned so as not to be able to see into cubicles when the doors were closed. Despite these reassurances, the pupils were unanimous that it was a huge invasion of their privacy, and many described feeling uncomfortable and upset by the presence of cameras:

> I really don't like the cameras in the toilets. When I first saw them I freaked. Me and Michelle were just like 'argh, what are we gonna do?' I've seen them at train stations and I get that, but in a school? Like what's gonna happen? (Sarah, UH)
>
> I try to spend as little time as possible in there now. You can't help but wonder if they are watching you. It's not just about going to the loo, it's

> blowing your nose [some members of the group laugh at this], yeah you know, putting your make up on and that. If girls are upset they run to the toilets so no-one sees them. That's why you do things like that in the toilets in the first place. (Lisa, UH)

The toilet represented a private domain for pupils, particularly females, in which they can discard their public 'performance' and reside 'backstage'.[13] It was considered to be a (relatively) private sanctuary in the midst of a public arena where intimate and emotional acts such as crying, or behaviours deemed unfit for public display such as blowing one's nose, could be conducted. It was regarded as an important area providing individuals with 'a chance to lay their masks aside'.[14] Even those pupils that perceived the CCTV cameras to have some positive attributes felt that their use in the pupil toilets was excessive:

> I agree with there being some cameras. My mum is a teacher at another school and her school is always getting vandalized. She wants more cameras to stop the school from being damaged and things being stolen ... I think that is fair enough where there is a real problem that might be solved with ... a few cameras. I don't like them being in the toilets. It does make me feel uncomfortable ... There are just some places that are private – and the toilet is one of them. (James, CC)

Other pupils did not like that the fact that the cameras were remote, since it removed the more conventional relationship between the watched and the watcher.

> It's just wrong. If they really wanted, then they could have a teacher there [...] to make sure everyone is behaving. If someone is watching a screen to watch us in the toilets, well then they might as well just be in the toilets. At least then you know that you're being watched. It's like being spied on. (Kelly, CC)
>
> Sometimes you forget. It's like being followed and you don't know. You think you're alone and you're not. [Pupil does a pretend shiver to signify that they find this creepy.] It's weird! (Milly, CC)

Overall, pupils at City Comprehensive and Urban High felt that the use of cameras in the toilet area was an unjustifiable invasion of their privacy, and disproportionate to any risk. Some pupils in England have refused to attend school citing CCTV in toilets as the reason and similarly, some parents resorted to removing their children from the school based on what they considered to be an 'outrageous invasion of privacy'.[15] This highlights that in some instances pupils do feel able to

DOI: 10.1057/9781137308863

mobilise themselves against situations in the school environment that they consider to be unacceptable.

Irrespective of the reasons why pupils dislike the use of CCTV in their school toilets, it is clear from the findings of this research that their presence makes them feel 'uncomfortable' and 'spied on' as well as undermining their privacy and fostering feelings of suspicion and mistrust. Pupils did not think the CCTV cameras were necessary. They did not identify any issues in the toilets to justify why CCTV would be a measured, rational and appropriate response. These two schools are not unique. As outlined earlier, some schools have cited the main objectives of CCTV in pupil's toilets to be as innocuous as to tackle the 'misuse of paper towels and soap', inhibit 'horseplay' or because 'wash basin areas were not being respected'.[16]

A 'chilling effect' and 'distanciation'

It was recognised that CCTV did not have the inherent power to evoke habituated anticipatory conformity amongst all pupils. Some would misbehave regardless of CCTV, whereas for others the presence of the cameras was enough to subdue interactions, whether they were mischievously inclined or not. Visual surveillance can have a 'chilling effect'[17] on associational activity, curtailing creativity, innovation and experimental modes of expression. This is of particular importance when used in institutions such as schools where creativity and intellectual exploration are central to education. CCTV cameras can render students 'acutely aware' that their actions are being recorded to the extent that some change their 'legitimate' behaviour 'due to a concern that their actions could be misinterpreted' by the watchers.[18] However, 50 per cent of teachers in the Association of Teachers and Lecturers (ATL) survey reported that they did not think that the pupils at their school behaved differently when they knew that they were being filmed by surveillance cameras.

Surveillance Schools are initiating a process of 'distanciation' whereby pupils are increasingly denied the opportunity for social interaction. Surveillance, whether facial recognition replacing registration, or fingerprinting to borrow library books, interrupts pupils' traditional patterns of 'sociation' or 'face-to-face' interaction with parents, teachers and their peers. In relation to CCTV it has been found that there is no longer any room for explanation or negotiation with teachers for issues such as late arrival; pupils can no longer buy their friend's lunch due to cashless catering systems; and 'automated text messages' impact on dialogue with parents about tardiness.[19]

DOI: 10.1057/9781137308863

Big Brother and the cult of the celebrity

Reality television shows and social networking sites featured prominently in discussions about CCTV and contributed towards the pupils' understanding of surveillance and privacy. It was common for the pupils to refer to the television show 'Big Brother' (Channel Four, UK). Series Six of the British show was being aired at the time that the fieldwork took place and so the link between the cameras in the schools and the Big Brother house, which was saturated with cameras, was immediately drawn: 'Yeah it's like we're on Big Brother! There are cameras everywhere you go' (Steph, UH).

Reality television shows such as 'Big Brother' create 'artificial realities' in which the contestants or participants are located in a specific environment,[20] whereas info-tainment shows (such as 'Caught on Camera' and 'Police, Camera, Action!') rely upon footage of real life situations and incidents. Both represent a shift towards identity as commodity, offering salacious insight into the private lives of others that would satisfy even the most demanding voyeuristic gaze. Privacy incursions are repackaged as entertainment and the blurring of boundaries regarding public and private activities contribute to the growing normalisation of (even desire for) being watched, devoured and consumed. The pupils equated the use of the cameras in the school with notions of celebrity and reality television although, importantly, they did this with humour. The idea that the mundane CCTV cameras of the school were transforming pupils into reality television stars generated amusement amongst some.

The audience was identified as an important factor in whether the presence of cameras would impact upon behaviour:

> (Milly, CC) I thought that, you know, when I was watching Big Brother. We all think 'oh my god' I can't believe they are doing that on the cameras, or you know, they're just playing up to the cameras, but then I think we're on the cameras all the time too. It gets you thinking about who is watching it.
>
> (Ben, CC) Yeah but they're on telly though – it's a bit different!
>
> (Milly, CC) Yeah of course it's different but you're still being filmed. It might not be being watched by millions of people. You won't end up being a celebrity like Nadia[21] [the majority of pupils in the focus group laugh at this] but there could still be someone watching you at any moment. Focused right in! I bet you wouldn't like that would you?!

The pupils in this focus group found the idea of being singled out by their school CCTV cameras as the main focus of attention as amusing and they gave the impression that they hadn't really considered this to be a possibility, nor had they entirely recognised that there might be somebody actively watching the cameras. Incidentally, it was this school, City Comprehensive, that had a member of staff who spent 'a large chunk' of his working day monitoring the cameras.

Demonstrating the resentment felt by some pupils about the use of CCTV in their school (and a flare for sarcasm), Alyssa from City Comprehensive stated: 'If I wanted to be on Big Brother I would have auditioned'. This illustrates that consenting, or rather 'wanting', something involves positive action, that is, in this case an audition, whereas the lack of consultation taken with pupils meant that no action could be taken to object to its implementation. This explicates the tacit consent that is often relied on by schools when implementing CCTV.

Avoidance and resistance

Pupils had developed tactics of resistance to the use of visual surveillance in school, categorised into three approaches: the avoidance of areas monitored by the CCTV cameras, restricting the ability of the CCTV cameras to identify individuals and repositioning the cameras.

Avoidance of areas monitored by the CCTV cameras

A number of pupils described how they, or other pupils at the school, tried to avoid the cameras by taking alternative routes around campus where possible, or congregating in areas where there was no camera coverage. As identified above, pupils were able to cite areas not covered by CCTV cameras – 'I know the places where there aren't any. Like I know that there aren't any on the stairs and I know that there aren't any where everyone goes and smokes' (Simon, UH). There was a shared knowledge amongst students about how to evade the omnipresent gaze of the cameras, particularly if deviant behaviour such as smoking was being contemplated, but at times the pupils just wanted to 'be themselves', free from the scrutiny of authority.

Restricting the ability of the CCTV cameras to identify individuals

Similar to the tactics used by intruders at Single Sex Comprehensive, pupils used clothing or barriers to evade the cameras.

DOI: 10.1057/9781137308863

> You can just put your your hood up if you don't want the cameras to see you. [Tim, UH]
>
> If you didn't want to be identified by the cameras then you could just cover your face. If you did that in the school though you're gonna attract attention. [Cait, UH]

Single Sex Comprehensive attempted to counter such tactics by introducing a colour-coded scheme for school ties. Each year group was allocated a different coloured tie, the idea being that even if a pupil hid their face, if they were still wearing their tie staff would be able to at least reduce the number of possible identities (obviously, this could be overcome by switching ties). Other strategies to attempt to delimit the ability of pupils to hide their identity include the ban of hooded jumpers and headgear on the school premises.

Repositioning the CCTV cameras

Pupils and teachers described how some cameras were tampered with to restrict vision, by tilting the camera to point up to the ceiling for example. More often than not, this was regarding as a 'prank' rather than to facilitate criminal activity, or even accidental.

> You can knock 'em out of the way, make them point upwards – especially the ones in the corridors 'cos the ceilings are so low. They are always spun round the wrong way. (James, CC)
>
> Some of the pupils knock the cameras so they are positioned incorrectly, just facing a wall. It's not always intentional. Sometimes they just get knocked when they are running down the corridors or putting their coats on. The ceilings in the corridors are quite low. You pick it up on the monitor soon enough and just reposition it again. (Site Manager, CC)

In another study of CCTV in schools, the researchers found that students would prevent the cameras from seeing them by putting 'chewy and stuff' over the cameras.[22] In the main, avoidance of, or interference with the cameras were taken to be harmless and humorous subversions of CCTV. Little malice or intention of criminal activity was detected, but rather a playful expression of rebellion that is common amongst young people. Interference with the cameras can be likened to relatively trivial activities performed by a boisterous minority in schools such as scrawling a name on a desk or experimenting with the boundaries of the school uniform.

Although the avoidance and resistance of the CCTV found in the schools are not regarded as serious attempts to defy the surveillance

DOI: 10.1057/9781137308863

practices and rules of the school, they do represent a clear challenge to surveillance. As such these acts cannot be ignored, irrespective of whether they represent a rejection of the CCTV per se, or the authority and disciplinary function of the school more widely. Several theorists have identified a general opposition to school authority amongst working-class adolescents in an attempt to maintain their own values and resist attempts by the dominant culture to impose theirs.[23] It is claimed that some pupils 'simply reject the overt and hidden curricula of the school' and resist as far as possible 'the covert teaching of punctuality, neatness, compliance and other more economically rooted values'.[24]

The case studies illustrate that pupils at the school believed that their right to privacy was being challenged by CCTV, and importantly this was considered to be evidence that they were not trusted. This is closely linked to the notion that there was a widespread perception of youth as problematic and essentially criminal in society and this transcended the school environment. Pupils claimed that their schools were like prisons and that they were being treated like criminals. Clearly, there is the potential for this to turn to resentment. Strategies of resistance to the surveillance cameras were identifiable amongst some pupils but these were regarded as playful anti-authority gestures that were not specifically retaliatory to the CCTV but rather the school itself. These themes are explored in greater detail in the following chapter.

Notes

1 ATL (2009): *85% of teachers have CCTV in their schools and nearly 25% worry about hidden cameras*. Available at: www.atl.org.uk/media-office/media-archive/cctv-in-schools.asp (accessed 04 September 2013).

2 Gill, M. and Loveday, K. (2003) 'What do offenders think about CCTV?', in M. Gill (ed.), *CCTV*. Leicester: Perpetuity Press, 81–92.

3 Taylor, E. (2010) 'Evaluating CCTV: why the findings are inconsistent, inconclusive and ultimately irrelevant', *Crime Prevention and Community Safety* 12: 209–32.

4 Repetto (1976) identified five types of displacement, with 'perpetrator displacement' later being added as an additional mode by Barr and Pease (1990). Repetto, T.A. (1976) 'Crime Prevention and the Displacement Phenomenon', *Crime and Delinquency* 22: 166–177; Barr, R. and Pease, K. (1990) 'Crime Placement, Displacement and Deflection', in M. Tonry and N. Morris (eds), *Crime and Justice: A Review of Research*. Chicago, IL: University of Chicago Press.

DOI: 10.1057/9781137308863

5 For example, see: Armitage, R., Smyth, G., and Pease, K. (1999) 'Burnley CCTV evaluation', in K. Painter and N. Tilley (eds), *Surveillance of Public Space: CCTV, Street Lighting and Crime Prevention.* New York: Criminal Justice Press, 225–49.

6 Waples, S., Gill, M., and Fisher, P. (2009) 'Does CCTV displace crime?', *Criminology and Criminal Justice* 9 (2): 207–24.

7 ATL (2009).

8 Graham, S. (1998) 'Towards the fifth Utility? On the extension and Normalisation of Public CCTV', in J. M. C. Norris, and G. Armstrong (eds), *Surveillance, Closed Circuit Television and Social Control.* Aldershot: Ashgate; Lyon, D. (2003) *Surveillance as Social Sorting.* London: Routledge.; Norris, C. and Armstrong, G. (1999) *The Maximum Surveillance Society: The Rise of CCTV*. Oxford: Berg.

9 Foucault, M. (1977) *Discipline and Punish: The Birth of a Prison*. London: Penguin Books, p. 201.

10 Ibid., 201.

11 Norris, C. and Armstrong, G. (1999), p. 2.

12 Furedi, F. (1997) *Culture of Fear: Risk Taking and the Morality of Low Expectation*. London: Cassell.

13 Goffman, E. (1959) *The Presentation of the Self in Everyday Life*. London: Penguin, p. 32.

14 Westin, A. (1967) *Privacy and Freedom*. New York: Antheneum, p. 35.

15 'School head defends toilets CCTV'. BBC News, 27 January 2009.

16 Ibid.

17 Bennett, C. and Raab, C. (2007) 'The privacy paradigm', in S.P. Hier and J. Greenberg (eds), *The Surveillance Studies Reader*. Berkshire: Open University Press.

18 Mccahill, M. and Finn, R. (2010), 'The social impact of surveillance in three UK schools: "angels", "devils" and "teen mums"', *Surveillance & Society* 7 (3/4): 273–89, p. 283.

19 Mccahill, M. and Finn, R. (2010).

20 Koskela, H. (2004) 'Webcams, TV shows and mobile phones: empowering exhibitionism', *Surveillance & Society*.2 (2/3): 199–215.

21 Nadia Almada was the winning contestant of the previous year's UK Big Brother series.

22 Mccahill, M. and Finn, R. (2010).

23 For example; Willis, P. (1977) *Learning to Labour: How Working Class Kids Get Working Class Jobs*. Farnborough: Saxon House; Corrigan, P. (1979) *Schooling the Smash Street Kids*. London: Macmillan Press.

24 Apple, M. (1982) *Education and Power*. London: Routledge, p. 96.

DOI: 10.1057/9781137308863

3

Lessons in Submission? The Societal Impacts of Surveillance Schools

Abstract: *Surveillance Schools undermine privacy, disrupt interactions premised on trust and in some instances even expedite the transition of pupils from the school to the prison. There are examples of student resistance to surveillance but this is responded to with exclusion, marginalisation and coercion.*

Keywords: trust, privacy, habituate, social networking, school-to-prison pipeline

Taylor, Emmeline. *Surveillance Schools: Security, Discipline and Control in Contemporary Education.* Basingstoke: Palgrave Macmillan, 2013.
DOI: 10.1057/9781137308863.

A common assumption prevails about technological school surveillance: that it simply automates existing processes and in doing so enhances security and relieves pressure on limited resource. It has been argued that 'wariness should be adopted in suggesting that [...] new technologies represent a radical change in surveillance practices'[1] since the apparatus may be 'functional equivalents'[2] of procedures in place hitherto. However, the findings from the handful of studies that have been conducted suggest that surveillance can radically change the educational environment.[3] Structural and technological surveillance can have implications and impacts that were not anticipated or desired and therefore cannot be presumed to be banal or neutral equivalents to what preceded them. The extent of the consequences are yet to be fully recognised, but there are clear indicators that the new modality of school surveillance is contributing to a reconfiguration of societal values such as privacy and trust, as well as undermining an already fractured belief in democratic participation amongst young people. This chapter identifies and explores a number of themes: privacy, mistrust, criminalisation, habituation to surveillance and resistance.

Privacy

Surveillance is often hailed as the nemesis of privacy, a tug of war between two social forces. But what is often left out of this fiercely fought battle are the questions 'why does privacy matter', 'what is actually being defended?' Isn't a claim to privacy old fashioned, antiquated, even embarrassing in the digital age? I argue that far from being obsolete, privacy is valued as much as ever, but we must reformulate our traditional perceptions of what it is.

Privacy is the most equivocal of all human rights in terms of definition and circumscription;[4] even its pronunciation is ambivalent (*pri*-vacy or *priv*-acy). The search for a universally accepted definition has produced a long and continuing debate 'that is often sterile, and, ultimately futile'.[5] However, the surveillance age makes it more pressing than ever to define what the term 'privacy' encompasses and stipulate what one's legitimate expectations of privacy can be. In its narrowest sense, privacy has been conceived as no more than a luxury enjoyed by the affluent in developed countries, and at its widest it has been espoused as the last opportunity for the weakest and poorest groups in society to defend themselves

DOI: 10.1057/9781137308863

against dominant power groups.[6] It has been conceptualised as providing a retreat where individuals can take refuge 'backstage'[7] and ultimately 'be themselves'.[8] Alan Westin provides a useful definition, albeit one that is unworkable on a legal basis, when he writes:

> Privacy is the claim of individuals, groups or institutions to determine for themselves when, how, and to what extent information about them is communicated to others. Viewed in terms of the relation of the individual to social participation, privacy is the voluntary and temporary withdrawal of a person from the general society through physical or psychological means, either in a state of solitude or small group intimacy or, when among larger groups, in a condition of anonymity or reserve.[9]

Privacy is claimed to facilitate many fundamental societal functions, such as personal autonomy and emotional release, as well as being integral to several aspects of social life, including but not limited to, psychological health,[10] ability to love[11] and the ability to form social relationships.[12] Privacy is an historical need that some commentators trace back to the beginning of humanity; it certainly predates the development of electronic surveillance. Furthermore, anthropological evidence suggests that most societies regard some domains of human activity as unsuitable for general observation, supporting the view that a need for privacy is near universal.[13]

A key issue with privacy, rendering it near impossible to enshrine definitively in law, is its highly subjective interpretation. Concerns 'vary over time, across jurisdictions, by different ethnic subgroups, by gender and so on'.[14] Privacy is further complicated by instances whereby it is beneficial to surrender one's personal data since 'complete absence from databanks is neither practical nor desirable' in the digital age.[15] But it is not just pragmatism that renders inclusion in the surveillance society desirable; increasingly the social world is digitally mediated and abstinence can be stultifying for relationship building in the surveillance society.

In the context of the school, privacy is perceptibly refracted. Schools are entitled to know certain personal details about the children for whom they are responsible, particularly when to withhold them could impact upon their welfare. For example, it is legitimate for schools to know about medical conditions and past educational attainment, although in the adult world this might be deemed as sensitive personal information to divulge to an organisation. It is understandable that privacy can lose

DOI: 10.1057/9781137308863

its significance when it conflicts with the potential of a child coming to harm. Brian Warnick, in his essay on the ethics of visual surveillance in schools, makes the point that 'having a bag searched for a weapon seems like it causes students little substantive harm…in comparison to the potential harm of being in a school with weapons'.[16] Safety versus privacy is always going to favour the former but of course that presumes that safety is in some way guaranteed and the surveillance is predicated on a rational and balanced perception of the risk.

Empirical research on the impact that surveillance in schools has on privacy has largely focused on the use of CCTV[17] but also internet usage and e-safety.[18] As outlined in the previous chapter, a recurrent theme throughout my research in three schools in the North of England was privacy. For students, the ability for CCTV to upset their expectations of privacy relied upon a number of factors such as location of the cameras, the rationale underpinning usage and whether they were being monitored continuously. The majority of students felt that a balance could be struck between surveillance and privacy in the school, but importantly most did not feel this had been achieved. A number of students were incensed about the lack of privacy they were afforded, not just in the school but more broadly as a result of surveillance technologies.

Discussion on privacy mainly focused on its use in the pupils' toilets at Urban High and City Comprehensive. Although some students were unable to articulate exactly how, they were firm that their notion of privacy had been compromised. As James, a pupil at City Comprehensive, stated: 'There are just some places that are private – and the toilet is one of them'. The toilet was representative of a private domain for pupils where they could, in Goffmanesque fashion, discard their public performance and reside backstage. Some pupils were particularly expressive about their need for (relative) privacy in the school environment and attributed a number of values to it, such as Lisa who went there to express emotions that she did not want publicly observed. The presence of CCTV in the toilets made pupils feel uncomfortable and spied upon, thereby eradicating its function as a private realm that permitted 'the restoration of the psyche and accumulation of the moral and emotional capital needed for the next emergence into public.'[19]

Social networking and privacy

The explosion of 'social networking sites' such as Facebook, YouTube, Bebo, MySpace and Twitter all offer extensive opportunity for the

DOI: 10.1057/9781137308863

individual to make public what was once private. Viewing individuals as solely passive targets of surveillance has become outdated. Individuals are increasingly active in revealing personal data about themselves and the expansion of social networking sites provide ample opportunity for them to do so. At times individuals seek to maximise their exposure rather than minimise it and do so by 'playing with, goading and yes, even flirting with surveillance'.[20] Social networking sites provide a kind of 'participatory panopticon' (see p. 83 for a description of the Panopticon).[21] The temptation is then to presume that young people no longer value, want, desire or need privacy, but the reality is far more complex. As the 'self' is mediated across an ever increasing number of forums, ownership of identity is actually becoming more important. Young people are certainly not strangers to surrendering information about themselves, but presuming that participation on social networking sites represents a full-scale relinquishing of autonomy, anonymity and privacy is absurdly simplistic and ignores the performativity involved in occupying online spaces.

Ervine Goffman explained how we carefully mediate ourselves in public in *The Presentation of the Self in Everyday Life*, and whilst it is perhaps more apt in the digital age to reframe this as 'The *Revelation* of the Self in Everyday Life', image management in the virtual domain is equally important, if not more so, than in face-to-face interaction. Facebook personas are just one of many representations of the self, carefully managed, 'self-fabrications'[22] which are most likely hyperbolic augmentations of how we wish others to 'see' us. Online performances are underpinned by 'careful and deliberate judgements about who sees what, as well as familiarity with both the technical and social tools at one's disposal'.[23]

The findings presented in Chapter 2 demonstrate that Westin's definition of privacy still resonates today. The key to privacy for the pupils wasn't the defacto withholding or revealing of information, but rather it was about determining for themselves when, how and to what extent information about them was communicated to others. Young people still wish to exert control over what information about them is publicly available, although what privacy is and the expectations that underpin it are being challenged. The virtual realm provides validation in a world increasingly fluid and uncertain. Zygmunt Bauman writes that 'the fear of disclosure has been stifled by the joy of being noticed'[24] but perhaps the greatest fear of all comes from the prospect of absenteeism. Threatened with being disregarded, we reassert our value through trading in the currency of 'likes', collecting 'friends' and attempting to

DOI: 10.1057/9781137308863

certify presence and mobility by 'checking in'. The social network doesn't recognise the existence of non-participants and so each transaction provides 'avidly sought proof of social recognition'.[25] A kind of 'I Tweet, therefore I am'.[26]

Privacy, both temporal and physical, has been transformed in the digital age, but the individuals who put forth their carefully scripted schemas still attempt to retain control over the boundaries of public and private. Image management is paramount. As boyd and Marwick[27] explain:

> All teens have a sense of privacy, although their definitions of privacy vary widely. Their practices in networked publics are shaped by their interpretation of the social situation, their attitudes towards privacy and publicity, and their ability to navigate the technological and social environment. As such they develop intricate strategies to achieve privacy goals. Their practices demonstrate privacy as a social norm that is achieved through a wide array of social practices configured by structural conditions. How teens approach privacy challenges the ways in which privacy is currently conceptualized, discussed, and regulated.

Young people often lack the skills and resources required to effectively express their desire and need for privacy, as well as the ability to negotiate with those who seek to control it. This becomes particularly problematic when the dominant discourse tramples the complex and subtle ways that young people construct privacy. Corporations are particularly conversant in pedalling the view that young people voluntarily relinquish all semblances of privacy.[28]

There is a need for further insight into the materialities of privacy as young people construct it, in both the virtual and physical realms to inform ethical debates on Surveillance Schools.

Mis-trust and the criminalisation of youth

As briefly explored in the previous chapter, privacy, surveillance and (mis)trust are all inextricably linked: 'privacy can expand only with trust, but trust can expand only with surveillance'[29] presenting an irredeemable paradox. As surveillance technology is increasingly relied upon, it can only ever produce a veneer of trust, or a 'thin trust'[30] which ultimately results in a need to delve deeper to extract confidence and verification, and thereby encroach further on expectations of privacy.

Henry A. Giroux claims that a 'deep distrust'[31] has enveloped interactions with children. Supporting this assertion the pupils were quick to correlate the increased use of surveillance at their schools with diminishing trust, as outlined in the previous chapter. References to the school being like a 'prison', and being treated like they 'were up to no good', demonstrates that some pupils perceived the increase of CCTV in schools was partly due to young people being regarded as inherently criminogenic. To explore this further, I accessed a number of online forums in which students discussed the surveillance practices being used in their school. Comparable sentiments were found in relation to both CCTV[32] and fingerprinting:[33]

> [CCTV] makes you feel like you're in a prison and that you can't be trusted. (Female, 11)
>
> I think that [fingerprinting] is stupid because it makes you feel like a criminal when you haven't done anything wrong [...] It's unfair and more adults steal books than children anyway! (Female, 14)
>
> I have not had my fingers printed and do not wish to. The people in our school who are most likely to steal a book aren't exactly the type to read! (Female, 13)

It is clear from the comments (and there are many more) that a large proportion of pupils equated surveillance with mistrust, and automatically linked its use to criminal activity. Of course there are students who believed the surveillance used in their schools was cool, futuristic and justifiable but this should not take away from the fact that many felt that their subjection to surveillance was criminalising them. Surveillance practices take on particular symbolic meanings in the school environment, conveying messages to young people about how they are viewed in society.

Moral panic: folk devils in hoodies and the neoliberal city

It is widely noted that 'national anxieties typically attach to youth' and 'moral panics have long targeted youth as the source of national troubles'.[34] The media perception of young people as highlighted by the pupils in the previous chapter contributes towards the criminalisation of a generation and cements stereotypical discourses of a dangerous and unruly youth in the public mind. Research has shown that the media overwhelmingly portrays young people as intrinsically criminogenic. For example, in the UK in 2008 more than half of the stories featuring teenage boys in national and regional newspapers were about crime.[35] Furthermore, the words most commonly used to describe them was 'yobs', 'thugs', 'sick' and 'feral'. The pernicious portrayal of youth in contemporary society has been identified in their media representation:

> When a photo of a group of perfectly ordinary lads standing around wearing hooded tops has become visual shorthand for urban menace, or even the breakdown of society, it's clear teenage boys have a serious image problem. The teen boys' "brand" has become toxic. Media coverage of boys is unrelentingly negative, focusing almost entirely on them as victims or perpetrators of crime – and our research shows that the media is helping make teenage boys fearful of each other.[36]

Moral panic about youth is nothing new, but in the neoliberal city this predilection becomes solidified in techniques of exclusion, marginalisation and criminalisation all of which are augmented by disproportionate surveillance. The intensification of urban social control can be attributed to the ascendance of global capitalism and the transformation of the urban economy from a public good to a corporately defined space. Young people find their access to regenerated spaces retracted via a number of exclusionary tactics that privilege the rights of those who engage in legitimised consumption activities. These include demonising means of expression and identity associated with young people, criminalising certain behaviours commonly engaged in by young people, as well as discriminating against their very presence in some areas. These avenues via which young people are excluded are outlined below.

Expression and style

The importance of dress and style as an expression of identity has been recognised as a key component of youth within sociological theory since the seminal work 'Folk Devils and Moral Panics' by Stanley Cohen in 1972. From the Mods and Rockers of the 1960s, the punk rockers of the 1970s to the Goths, Chavs and Emos in contemporary Britain, the myriad subcultures have all expressed themselves through clothes, hairstyles and music. As has happened previously with various youth cultures, there has been a societal backlash against those that are associated with a certain form of youth culture. Most recently it is the metonym 'hoodie' that has been vilified as the modern day folk devil representing a feral and degenerate youth. The term is often used as shorthand to denote a debauched generation and the havoc wreaked on society as illustrated in newspaper

DOI: 10.1057/9781137308863

headlines such as *'Hoodie Hell on Streets Every 8 Secs'*[37] and *'Hoodie Allowed to Break ASBO 16 Times Before Finally Being Sent To Prison'*.[38]

Resulting from its symbolic representation of delinquency and social disorder, the hooded top has undergone a widespread interdiction from shopping centres, pubs, clubs, schools and supermarkets in the UK.[39] Many schools have also implemented the ban on hooded tops. One of the key reasons appears to be that despite efficacious claims about CCTV, it is rendered largely ineffectual when individuals wear hooded tops or any form of headwear that hides their identity. As one teacher stated, the bans were not so much about crime control as the control of youth expression:

> You can't wear hooded tops and caps in our school. I don't agree with the ban but the argument was crime could be committed in the school and they could remain anonymous on the CCTV cameras. The subtext is they are trying to stop all youths wearing things they consider offensive. It's a scare tactic.[40]

The vilification of clothing associated with young people infringes upon cultural expression and advances a discourse of a dangerous and criminogenic youth. A similar 'war on youth culture' has been identified in US schools following the shooting at Columbine High School, Colorado on 20 April 1999. Following the event, a societal backlash against 'Goth' subculture ensued in which the symbolic representations of social ills were recast as the cause of them.[41] The political economy of the Surveillance School focuses on the creation of disciplined hegemonic consumers and those perceived as challenging such imperatives through oppositional cultural styles and praxis will not surprisingly find themselves under increased scrutiny supported by a raft of surveillance measures.

Behaviour and activities

The burgeoning of surveillance is largely perceived as requisite to contain an increasingly savage youth spawned from the generic moral decay apparent in wider society. In a similar vein to the demonising of certain styles of dress associated with young people, certain behaviours have suffered from being reclassified as deviant. The reclamation of public urban space by private entities is a paradoxical process: whilst culture and diversity are espoused in political rhetoric as something to be celebrated and treasured for enriching the life of the urbanite, certain forms of cultures associated with some groups are actively oppressed and outlawed. The

DOI: 10.1057/9781137308863

criminalisation of skateboarding in major cities such as Manchester and Liverpool in England is just one example. Roy Coleman remarks how CCTV becomes complicit in the excessive control of young people in urban spaces: 'cameras aid the strategic balance between aesthetics and function' whereby 'any notion of the city as a space of cultural expression for younger people continues to be highly circumscribed'.[42] The construction of activities such as skateboarding, which is not intrinsically deviant or harmful, as criminal belies the neutrality of urban social control, and instead highlights how its operation reflects the interests of controlling agents, increasingly with corporate rather than civic interests.

Young people have been excluded from urban centres on grounds of their cultural praxis, behaviour and style. To a certain extent, this exclusion still presents a degree of choice amongst young people in that they could choose not to wear hooded tops, or they could not partake in activities that are perceived as a nuisance such as skateboarding or 'hanging around'. But perhaps the most insidious developments are those that target and exclude youth based unequivocally on age alone. For example, marketed as the 'ultrasonic teenage deterrent' by its vendor Compound Security Systems Ltd., The Mosquito™ emits a high pitch sound that is audible only to young people due to a naturally occurring phenomenon known as presbycusis (age-related hearing loss). The high pitch sound is thought to be uncomfortable and irritating for young people to listen to, and so it is aimed at dispersing unwanted gatherings of youths and teenagers. It has been implemented in many contexts such as shopping malls, public transport, schools and so on.

School-to-prison-pipeline

The use of surveillance, both technological and structural, as well as the trend to outsource school discipline arguably fast-tracks some students along the route to prison, creating what John Devine terms a 'school-to-prison pipeline'[43] and Nolan and Anyon, a 'school/prison continuum'.[44] The reliance on more severe consequences in response to student disruption has resulted in an increase of referrals to the juvenile justice system for infractions that were once handled in school. Rule breaking is redefined as a criminal matter and troublesome students are criminalised, symbolically if not legally, and exposed to the criminal justice system from an early age.[45] Zero tolerance is particularly pernicious as there is

DOI: 10.1057/9781137308863

strong correlation between suspension/expulsion from school and entry into the criminal justice system.[46] Otherwise well-behaved children are being funnelled into the criminal justice system for egregious decisions made about them whilst they were in school. In England, half of young offenders in custody had been excluded at some point during their school years. The use of zero tolerance coupled with onsite police means that trivial forms of student misconduct that were once handled informally by teachers and school administrators now result in police arrest.[47]

Increasingly comparisons are made between schools and prisons[48] and some go so far as to see the school as 'a subsector of the larger prison-industrial complex'.[49] Some schools are located in disused prisons, whereas others are actively building estates with designs akin to prison facilities. As an example of such 'penal swarming',[50] a school in the West Midlands caused outrage when it released plans to build a 'detention wing' two miles from its main campus, where up to 12 'troublemakers' would be taught under strict conditions.[51] Under the scheme, the pupils are to be 'rewarded' for good behaviour with playtime and be allowed to rejoin their classmates if they prove themselves to be 'rehabilitated'. The school had already caused concerns when it permanently expelled 15 students upon becoming an academy in 2009, resulting in it having the second highest expulsion rate in England.

Resistance

Several theorists have identified a general opposition to school authority amongst working-class adolescents in an effort to maintain their own values and resist attempts by the dominant culture to impose theirs.[52] Michael Apple identified that some pupils 'simply reject the overt and hidden curricula of the school' and resist as far as possible 'the covert teaching of punctuality, neatness, compliance and other more economically rooted values'.[53] More specifically a range of techniques of resistance to school surveillance has been identified by scholars.

In my research on the use of CCTV in schools I found avoidance of or interference with CCTV cameras in schools were largely interpreted as harmless and humorous subversions of the operation of visual surveillance within the school. Andrew Hope reports similar findings in relation to internet surveillance in schools in which little malice or intention of criminal activity was apparent in the pupil's attempts to access prohibited sites.[54] As Weiss asserts, 'everyday surveillance is matched by everyday resistance'[55] and so perhaps the routinised nature of both surveillance and resistance defuses its gravitas. There have, however, been more serious attempts to undermine surveillance systems in schools, from coordinated walk-outs in response to CCTV in classrooms, to court battles challenging the use of RFID (see Case Study Box).

CASE STUDY: RFID – 'THE MARK OF THE BEAST'?[56]

The most sustained and fervent resistance to RFID has come from perhaps an unlikely source, Biblical prophecies. In chapter 13 of The Book of Revelation, the last book of the New Testament outlines:

> *And he causeth all, both small and great, rich and poor, free and bond, to receive a mark in their right hand, or in their foreheads* (Rev. 13:16–17)
>
> *And that no man might buy or sell, save he that had the mark, or the name of the beast, or the number of his name.* (Rev. 13:16–17)

This has been interpreted by some evangelical Christian groups as relating to the use of RFID in the retail sector. The reference to being able to 'buy or sell' precluding those that do not have the RFID chip upon them, or even embedded under their skin. More broadly all applications of RFID, including in schools, have been interpreted by some as the 'mark of the beast'.

John Jay High School in San Antonio, Texas introduced RFID tags for all students as part of their Student Locator Project in 2012. Fifteen-year-old student Andrea Hernandez refused to wear the RFID tag on grounds that it contravened her religious beliefs and was promptly suspended by the school. The school proposed a deal – if Andrea and her family would agree to cease criticising the initiative they would provide her with a chip-less badge, but her father argued that this would still amount to tacit consent for the overall programme. The case went to court.

The judge ruled that Hernandez's refusal to wear the badge without the RFID chip undermined her claim that the school district was violating her religious freedom stating: 'Plaintiff's objection to wearing the Smart ID badge without a chip is clearly a secular choice, rather than a religious concern.'[57]

The judge concluded that the student must either accept the school's compromise and wear the badge without the chip or move to a different school. However, this is particularly problematic since Northside Independent School District (NISD) in which her school is located is planning on rolling out the $500,000 Student Locator Project to all of its 112 schools reaching a student population of 100,000. A non-profit civil rights law group is intending to appeal the judgement on grounds that the judge may not question the validity of an individual's religious beliefs.

There are also clever ways in which young people attempt to highlight the absurdity of some school surveillance systems; the online guides demonstrating how to circumvent AFIS by making fake fingerprints from melted down 'gummy bears' (or other gelatinous sweets) spring to mind. However, for some the power of surveillance is able to disrupt and silence any serious contention or dissent.

Jennifer Weiss draws upon one of the largest recorded protests in response to surveillance practices in schools to explore the role of resistance.[58] In the autumn of 2005, 1,500 students at a large comprehensive school in the Bronx, New York staged a walkout in protest to the implementation of metal detectors and security cameras at their school. The students marched three miles under police escort to demand a meeting with the region's superintendent and other Department of Education officials. Despite generating a huge amount of public and media interest, the protest was not effective in reversing the decision to install metal detectors at the school; in fact Weiss reports that it actually increased them.

As the students in the Bronx discovered, direct action doesn't always result in young people being heard, and resistance is not without consequence. For example in Brazil in 2012, 107 students were suspended from school after they protested about the installation of cameras which they claimed had been introduced with a lack of dialogue between school administrators and the pupils. Similarly, pupils opting out of fingerprinting initiatives have found that they couldn't purchase lunch at the school canteen since the payment system was underpinned by biometrics, couldn't borrow library books, couldn't vote for homecoming King and Queen and couldn't even enter certain classrooms. Compliance with surveillance is therefore based on coercion. Students who assert their

DOI: 10.1057/9781137308863

rights are met by schools with an arsenal of disciplinary mechanisms to stifle dissent. The real tragedy is when young people find that as a result of voicing concerns or mobilising their democratic rights, they are then subjected to penalties or sanctions. In Chapter 2, it was noted that police and security personnel are increasingly involved in profiling students. Oppositional behaviour and resistance is possibly the surest way for students to ensure that they capture the attention of those compiling threat assessments. The most depressing outcome would be for students to internalise the futility of questioning authority and concede to conformity.

Following the walk-out in the Bronx, the students developed alternative means to avoid and evade surveillance, including the 'tactical avoidance' of interaction with security officers. In this way resistance became 'masked in a kind of complicity'[59] since it would not be easily observable from the outside, but still represented a challenge to those in authority. Students might lack the power needed to change policy but they retain the ability to 'cope with, negotiate, and respond to these practices and injustices'.[60] Whilst the Surveillance School seeks to control pupils, it also provides the space to develop identities, experiment with boundaries and resist total compliance.

It is claimed that 'students have historically found spaces in school where they can avoid staff surveillance.'[61] This is not always with the intention to misbehave, but sometimes, students require 'time out' and the ability to withdraw from pressured social interactions and the trials and tribulations of adolescence played out publically in the school. However, with the use of CCTV and RFID these spaces might well be disappearing.

Developmentally the full impact of surveillance on young people is yet to be discovered but it is clear that it erodes belief in democratic systems, teaches young people that the only means of trust is through technological means, and takes away privacy. There are clearly going to be implications for the structural and institutional disdain that the youth of today are treated with.

Notes

1 Hope, A. (2010) 'Student resistance to the surveillance curriculum', *International Studies in Sociology of Education* 20 (4): 319–34, p. 322.

DOI: 10.1057/9781137308863

2 Feeley, M. (2003) 'Crime, social order and the rise of neo-Conservative politics', *Theoretical Criminology* 7 (1): 111–30, p. 118.
3 For example, see; Gallagher, M. (2010) 'Are schools Panoptic?', *Surveillance & Society* 7 (3/4): 262–72; Kupchik, A. (2010) *Homeroom Security: School Discipline in An Age of Fear*. New York: New York University Press; Mccahill, M. and Finn, R. (2010), 'The social impact of surveillance in three UK schools: "angels", "devils" and "teen mums"', *Surveillance & Society* 7 (3/4): 273–89; Taylor, E. (2010) 'I spy with my little eye: the use of CCTV in schools and the impact on privacy', *Sociological Review* 58 (3): 381–405; Weiss, J., (2008) *Under the Radar: School Surveillance and Youth Resistance*. The City University of New York.
4 Michael, J. (1994) *Privacy and Human Rights*. Hampshire: Dartmout Publishing Company Ltd.
5 Wacks, R. (1980) *The Protection of Privacy*. London: Sweet & Maxwell, p. 10.
6 Michael, J. (1994).
7 Goffman, E. (1959) *The Presentation of the Self in Everyday Life*. London: Penguin, p. 32.
8 Robertson, A.H. (1973) *Privacy and Human Rights*. Manchester: Manchester University Press, p. 3.
9 Westin, A. (1967) *Privacy and Freedom*. New York: Antheneum, p. 7.
10 See for example; Ingham, R. (1978) 'Privacy and psychology', in J. Young (ed.), *Privacy*. Chichester: Wiley and Sons.
11 See for example; Fried, C. (1971) *An Anatomy of Values: Problems of Personal and Social Choices*. Cambridge: Cambridge University Press.
12 See for example; Gerstein, R. (1984) 'Intimacy and privacy', in F. Schoeman (eds), *Philosophical Dimensions Privacy: An Anthology*. Cambridge: Cambridge University Press.
13 Michael, J. (1994).
14 Bennett, C. and Raab, C. (2007) 'The privacy paradigm', in S.P. Hier and J. Greenberg (eds), *The Surveillance Studies Reader*. Berkshire: Open University Press, p. 340.
15 Stalder, F. (2002) 'Opinion: Privacy is not the antidote to surveillance', *Surveillance & Society* 1 (1): 120–24, p. 122.
16 Warnick, R. (2010) 'Surveillance cameras in schools: an ethical analysis', *Harvard Educational Review* 77 (3): 317–43, p. 323.
17 For example, see; Mccahill, M. and Finn, R. (2010); Taylor, E. (2010).
18 For example, see: Barnard-Wills, D. (2012) 'E-safety education: young people, surveillance and responsibility', *Criminology and Criminal Justice* 12 (3): 239–55; Hope, A. (2008) 'Internet pollution discourses, exclusionary practices and the "culture of over-blocking" within UK schools', *Technology, Pedagogy and Education* 17 (2): 103–13.

DOI: 10.1057/9781137308863

19 Sykes, C.J. (1999) *The End of Privacy: The Attack on Personal Rights – at Home, at Work, on-Line and in Court*. New York: St Martin's Griffin, p. 13.
20 Groombridge, N. (2002) 'Crime control or crime culture TV?', *Surveillance & Society* 1 (1): 30–46, p. 43.
21 Whitaker, R. (1999) *The End of Privacy: How Total Surveillance is Becoming a Reality*. New York: New Press, p. 139.
22 Bauman, Z. and Lyon, D. (2013) *Liquid Surveillance*. Cambridge: Polity Press, p. 34.
23 Steeves, V. (2012) 'Hide and seek: surveillance of young people on the internet', in D. Lyon, K.D. Haggerty, and K. Ball (eds), *The International Handbook of Surveillance Studies*. Oxon: Routledge, 352–60, p. 356.
24 Bauman, Z. and Lyon, D. (2013), p. 23.
25 Ibid., p. 24.
26 Orenstein, P., 'I Tweet, therefore I am', *The New York Times*, 30 July 2010.
27 boyd, D. and Marnick, A. (2011) 'Social steganography: privacy in networked publics', *Paper presented at ICA on* 28 May 2011, Boston, MA.
28 Steeves, V. (2012).
29 Ericson, V. and Haggerty, K.D. (1997) *Policing the Risk Society*. Oxford: Oxford University Press, p. 117.
30 Crawford, A. (2000) 'Situational crime prevention, urban governance and trust relations', in A. Von Hirsch, D. Garland, and A. Wakefield (eds), *Ethical and Social Perspectives on Situational Crime Prevention* Oxford: Hart Publishing, p. 209.
31 Giroux, H.A. (2003) 'Racial injustice and disposable youth in the age of zero tolerance', *International Journal of Qualitative Studies in Education* 16 (4): 553–65, p. 554.
32 CBBC 'Does CCTV make you feel safer?', *Newsround*, 1 June 2007. Available at: http://news.bbc.co.uk/cbbcnews/hi/newsid_6700000/newsid_6700500/6700513.stm (accessed 08 April 2013).
33 CBBC 'Should kids get fingerprinted?', *BBC Newsround*, 25 July 2002. http://news.bbc.co.uk/cbbcnews/hi/chat/your_comments/newsid_2146000/2146389.stm (accessed 08 April 2013).
34 Ruck, M., Harris, A., Fine, M., and Freudenberg, N. (2008) 'Youth experiences of surveillance: a cross-national analysis', in M. Flynn and D.C. Brotherton (eds), *Globalizing the Streets: Cross-Cultural Perspectives on Youth, Social Control, and Empowerment*. New York: Columbia University Press, 15–30, p. 16.
35 Echo (2009) 'Teenage boys and the media', *www.womeninjournalism.co.uk* (accessed: 09 April 2013).
36 Bawden cited in Garner, R., '"Hoodies, louts, scum": how media demonises teenagers', *The Independent*, 13 March 2009.
37 Lyons, J., 'Hoodie hell on streets every 8 secs', *Mirror*, 4 February 2009.

DOI: 10.1057/9781137308863

38 'Hoodie Allowed to Break ASBO 16 Times Before Finally Being Sent To Prison', *Mail Online*, 6 November 2007.
39 For example, see Coleman, R. (2004) *Reclaiming the Streets: Surveillance, Social Control and the City*. Devon: Willan publishing.
40 Quoted in Barkham, P., 'How a top can turn a teen into a hoodlum', *The Guardian*, 14 May 2005.
41 Muzzati, S.L. (2004) 'Criminalising marginality and resistance: Marilyn Manson, Columbine and cultural criminology', in J. Ferrell, et al. (eds), *Cultural Criminology Unleashed.* London: Glasshouse Press, 143–54.
42 Coleman, R. (2004), p. 187.
43 Devine, J. (1996) *Maximum Security: The Culture of Violence in Inner-City Schools*. Chicago: University of Chicago Press.
44 Nolan, K. and Anyon, J. (2004) 'Learning to do time: willis' cultural reproduction model in an era of deindustrialization, globalization, and the mass incarceration of people of color', in N. Dolby, G. Dimitriadis, and P. Willis (eds), *Learning to Labor in New Times.* New York: Routledge, 114–29, p. 114.
45 Hirschfield, P.J. (2008) 'Preparing for prison? The criminalization of school discipline in the USA', *Theoretical Criminology* 12: 79–101.
46 Skiba, R., Reynolds, C.L.R., Graham, S., Sheras, P., Conoley, J.C., and Garcia-Vazquez, E. (2006) 'Are zero tolerance policies effective in the schools? An evidentiary review and recommendations', American Psychological Association Zero Tolerance Task Force.
47 Beger, R.R. (2002).; Dohrn, B. (2001) '"Look out kid/It's something you did": zero tolerance for children', in W. Ayers, B. Dohrn, and E. Ayers (eds), *Zero Tolerance: Resisting the Drive for Punishment in Our Schools.* New York: The New Press, 89–113.; Fuentes, A. (2013) *Lockdown High: When the Schoolhouse Becomes a Jailhouse*. London: Verso.
48 Parenti, C. (2000) *Lockdown America: Police and Prisons in the Age of Crisis.* New York: Verso; Giroux, H.A. (2003); Nolan, K. and Anyon, J. (2004).
49 Lewis, T. (2003) 'The surveillance economy of post-Columbine schools', *Review of Education, Pedagogy, and Cultural Studies* 25 (4): 335–55, p. 36.
50 Simon, J. (2007) *Governing Through Crime: How the War on Crime Transformed American Democracy and Created a Culture of Fear*. Oxford: Oxford University Press, p. 225.
51 Duel, M., 'Fury as school with second-highest expulsion rate in England builds "prison block" to house 12 troublemaker pupils', *The Daily Mail*, 10 September 2012.
52 For example, see Corrigan, P. (1979) *Schooling the Smash Street Kids*. London: Macmillan Press; Willis, P. (1977) *Learning to Labour: How Working Class Kids Get Working Class Jobs*. Farnborough: Saxon House.
53 Apple, M. (1982) *Education and Power*. London: Routledge, p. 96.

DOI: 10.1057/9781137308863

54 Hope, A. (2005) 'Panopticonism, play and the resistance of surveillance: case studies of the observation of student Internet use in UK schools', *British Journal of Sociology of Education* 26 (3): 359–73.

55 Weiss, J. (2008), p. 595.

56 Kaliski, B.S. (no date) 'Thoughts on Radio Frequency Identification (RFID) Technology and Revelation's "Mark of the Beast"', http://www.infosecwriters.com/hhworld/hh10/rfid.htm (accessed 19 September 2013).

57 *A.H. V. Northside Isd, Dkt. No. Sa-12-Ca1113-Og (W.D. Tex. January 8, 2013).*

58 Weiss, J. (2011) 'Valuing youth resistance before and after public protest', *International Journal of Qualitative Studies in Education* 24 (5): 595–99.

59 Weiss, J. (2008), p. 9.

60 Weiss, J. (2011), p. 595.

61 Hope, A. (2005), p. 371.

DOI: 10.1057/9781137308863

4
Panoptic Pedagogy and the Political Economy of Surveillance Schools

Abstract: *Discipline has always been a central concern of schools and this has led many scholars to employ the Panopticon as a useful metaphor for understanding the distribution and exercise of power in educational institutions. This chapter will consider the extent to which schools are panoptic. Following on from this discussion, the way in which fear fuels the surveillance economy is explored. It is argued that the societies in which Surveillance Schools are located are increasingly shaped by the neoliberal regimen contributing towards new pedagogies based on risk and corporate concerns.*

Keywords: discipline, Panopticon, neoliberalism, media

Taylor, Emmeline. *Surveillance Schools: Security, Discipline and Control in Contemporary Education.* Basingstoke: Palgrave Macmillan, 2013.
DOI: 10.1057/9781137308863.

It is difficult to know where to start writing about the aetiology of Surveillance Schools. Their emergence is symptomatic of broad societal, economic, technological and political trends, the corollaries of which manifest themselves in school policies and practices oriented along an axis of risk, fear, corporatisation and punitive pedagogies. As outlined in the Introduction, schools have always been about surveillance, just as they have always sought to socialise pupils into the dominant culture of society. Part of this process is ensuring that young people are disciplined and ready to conform to the requirements of the labour force and the market; this is nothing new. But *something* new has emerged in schools since the 1990s, gathering momentum in the early years of the twenty-first century. Schools are very different places today than they were a little over 20 years ago.

The chapter begins by explicating discipline as an integral function of the school. School discipline has been interpreted as both emancipatory[1] and oppressive[2] but it is recognised that within surveillance studies it has been irrevocably anchored to the anatomy of power embodied by the Panopticon. Indeed analysis of surveillance 'cannot evade some interaction with the panopticon'.[3] As such, this chapter examines whether schools can be considered panoptic. The second part of this chapter shifts focus from the modern industrial age to the global post-industrial society in which schools are now located to explore the criminalisation and corporatisation of schools. It cannot be ignored that neoliberalism is a 'major fact of our time'[4] and this has vast implications for the way society and its institutions are organised and governed.

It is worthwhile noting that the accounts offered are articulated to different degrees across national contexts. Furthermore, they are interlinked, overlapping and tortuous; attempting to impose a linear narrative on the emergence of Surveillance Schools is simply a heuristic device to explicate them judiciously.

The spirit of discipline and the role of schools

The dawn of the twentieth century brought with it radical changes in the way society was structured and organised. The economy was transformed through industrialisation, the emergence of large corporations and a shift from a producer-based to a consumer-based market.[5] Urbanisation and immigration reconfigured the social landscape as workers moved

DOI: 10.1057/9781137308863

from organic communities in which they lived and worked to the city. Modernity brought with it a radically mechanistic and individualised perspective that challenged the moral and civic cohesion necessary for collective identity and shared public life. The school would become a key institution for socialisation and the vehicle through which social beings could adapt to the rapidly changing world. As Emile Durkheim postulates, education 'is above all the means by which society perpetually recreates the conditions of its very existence',[6] or in Althusserian parlance, education is the process through which society makes the individual in its own image. The school works to harness and invest in its pupils shared morals, practices and ideals. Despite the plurality of culture, there 'exists a certain number of principles which, implicitly or explicitly, are common to all'.[7]

In the modern industrialist nation, the school prepared its students for their roles in the industrial workforce.[8] Children learnt the behavioural values of the dominant culture and the skills required to facilitate a smooth transition from the classroom to the factory floor. As Michel Foucault describes, the school was a 'mechanism for training', 'a pedagogical machine', the product of which was an ordered and compliant workforce.[9] A 'spirit of discipline'[10] in an institution focused on 'learning to labour' was necessitated to achieve subjugation and compliance.

Aside from serving social functions, education was fundamentally emancipatory according to the early education scholars. For Emile Durkheim, the discipline encapsulated in the modern school had a 'double objective'; it promoted 'a certain regularity in people's conduct' that was essential for a cohesive society, but it also provided individuals 'with determinate goals' that served to 'limit their horizons'.[11] Discipline in this respect essentially places restrictions on unabashed hedonism, unrealistic aspirations and desire, which according to Durkheim, 'teaches us to restrain and master ourselves' which 'is a means of emancipation and of freedom'.[12] At the opposite end of the spectrum, critical scholars argue that discipline in schools is oppressive and meritocracy is a charade masking the true structures of capitalist society. From this perspective 'since its inception [...] the public school system has been seen as a method of disciplining children in the interests of producing a properly subordinate population'.[13] Whether emancipatory or oppressive, the school must ensure that it produces 'useful individuals' required for societal functioning.[14]

DOI: 10.1057/9781137308863

In English elementary schools, discipline was regarded as a process of obedience training, as this excerpt written in 1895 conveys:

> By disciplinary influence in the narrow school sense, is meant the combined agencies which the teacher brings to bear upon the child in order to make him [sic] amenable to law and order, and to arouse such energy as he possesses in a way to induce him voluntarily to put forth his efforts in the direction we wish; to train him to steady application and prompt and willing obedience, whilst at the same time we strengthen him to act more and more independently until he is able to become a law unto himself.[15]

Socialisation in this respect is distinct from coerced conformity. The role of discipline is to encourage children to voluntarily behave in ways that are deemed appropriate. The ultimate outcome is for children to possess 'a responsible regard for authority whilst also maintaining an intelligently flexible attitude towards established practices'.[16]

Foucault and discipline

Recent decades have seen theorising about discipline become preoccupied with the mechanism of power extolled by Michel Foucault in his seminal text *Surveiller et Punir* (1977), or *Discipline and Punish* as it is widely known by its English translation. The text has become a superlative touchstone for surveillance studies, and whether scholars agree or disagree with its applicability, it has become impossible to side-step its prominence in the field. There are several theories that are relevant to the school. Part Three, entitled 'Discipline', is organised into three sections, respectively 'docile bodies', 'the means of correct training', and 'panopticonism'. It is within these sections that Foucault elaborates his account of a new anatomy of power that emerged during the seventeenth and eighteenth centuries:

> 'Discipline' [...] is a type of power, a modality for its exercise, comprising a whole set of instruments, techniques, procedures, levels of application, targets; it is a 'physics' or an 'anatomy' of power, a technology.[17]

With functionalities resonant of Althusser's 'ideological state apparatuses' and Goffman's 'total institutions', Foucault identified 'complete and austere institutions'[18] (Fr. *Institution disciplinaire*) as the origin of discipline. The school is just one manifestation of the 'enclosed institutions of discipline' (alongside hospitals, prisons, asylums).[19] Foucault identifies

DOI: 10.1057/9781137308863

these disciplinary mechanisms 'at work in secondary education at a very early date, later in primary schools' before 'they slowly invested the space of the hospital; and in a few decades, they restructured the military organisation'.[20]

Key to Foucault's theory of discipline is that it creates new forms of individuality for bodies which enables them to perform within the constraints of societal norms and regulations. The individuality which discipline constructs has four characteristics: it is cellular (determining the spatial distribution of bodies), it is organic (it codes activities as normal for the bodies performing them), it is genetic (by controlling the accumulation of time) and it is combinatory (through the accumulation of forces). As Foucault articulates, discipline 'operates four great techniques; it draws up tables; it prescribes movements; it imposes exercises; lastly, in order to obtain the combination of forces, it arranges "tactics"'.[21] The apparatus and formal procedures of the school are intrinsic in this process: the school uniform depersonalises, the school bell signals when to start things and when to end, the timetable regulates the passing of time and the subject offerings ensure that pupils are the recipients of what others consider to be important knowledge. And so discipline comes to be 'utilised by schools that find in it a means of reinforcing or reorganising their internal mechanisms of power'.[22] Foucault's analysis of disciplinary power was exemplified in architectural form in the Panopticon, and it is this concept that has interminably occupied surveillance scholars.

Discipline and the Panopticon

It cannot be ignored that a fundamental strand of the sociological discourse on surveillance, and in particular CCTV, has been based on the consideration, if not direct application, of the Panopticon. Indeed, it has become one of 'the most powerful metaphors'[23] for explicating the theoretical and social significance of surveillance in contemporary society.

The Panopticon (meaning 'all-seeing'), was a design for a model prison drawn up by Samuel Bentham in the late eighteenth century. His brother, Jeremy Bentham, became somewhat transfixed by the design and set about trying to have the architectural blueprint realised. His attempts to convince the government to support the project were captured in a series of letters in which he extolled the ability of the design to reform morals, invigorate industry and preserve health, amongst other attributes.[24]

In its simplest form, the Panopticon was a building of circular design with a central observation tower that permitted the constant

DOI: 10.1057/9781137308863

surveillance of inmates by an unseen observer. Control was maintained by the constant unawareness of whether one was under surveillance at any given time, but, obliged to assume they were, conformity ensued. The Panopticon rendered all inmates permanently visible and thereby expedited intervention to any displays of non-conformity. Over time, those subjected to this field of vision would concede to anticipatory conformity resulting in the automatic functioning of power.

The architectural design of the Panopticon inspired in Foucault an understanding of an anatomy of power which he termed 'panopticonism'. He saw in it 'a mechanism of power reduced to its ideal form'.[25] It is important to make the distinction between the Panopticon as an intended structure comprised of bricks and mortar and its adjective derivatives which operate at a more 'abstract and superordinate level'.[26] For Foucault, the ability to watch inscribed in the structures of the Panopticon represented the shift from coercive means of power to disciplinary mechanisms of control:

> There is no need for arms, physical violence, material constraints. Just a gaze. An inspecting gaze, a gaze which each individual under its weight will end by internalising to the point that he is his own overseer, each individual thus exercising this surveillance over and against himself. A superb formula: power exercised continuously and for what turns out to be minimal cost.[27]

Whereas in Orwell's Oceania the 'telescreens' and imposing posters of Big Brother act as a constant reminder of the totalitarian regime and the requirement of compliance amongst its citizens, panopticonism renders these visual *aide memoires* obsolete. Power is not imposed upon individuals by the crude use of force, coercion or prohibition, but rather it is internalised and becomes transmitted by and through them. Panopticonism induces 'a state of consciousness and permanent visibility that assures the automatic functioning of power'.[28] Foucault theorised that the mechanisms of panopticism have a tendency to become 'de-institutionalized', that is 'to emerge from the closed fortresses in which they once functioned and to circulate in a 'free' state'[29] eventually becoming the model of power for the whole of society.

Surveillance Schools and the Panopticon

It has become increasingly commonplace for scholars to refer to schools as panoptic.[30] Whilst upon first inspection schools can appear to be

DOI: 10.1057/9781137308863

'disturbingly panoptic',[31] or as an 'empire of the gaze',[32] the interpretation of panopticonism requires careful consideration if it is to retain any value as a theoretical article. Michael Gallagher asserts that although it is 'tempting' to draw parallels, the school cannot be considered panoptic due to the surveillance being discontinuous rather than total (therefore open to resistance and evasion).[33] However, in Foucault's panopticonism the need for surveillance is in fact decremental, as he states 'the surveillance is permanent in its effects, even if it is discontinuous in its action' and 'the perfection of power should tend to render its actual exercise unnecessary'.[34] Furthermore, Gallagher's empirical study makes no mention of electronic surveillance techniques in use at the primary schools that he studied. When viewed as an ideal type, panopticonism is no longer a question of absolutes but rather one of intensity and opacity. It is useful to conceive of contexts along a panoptic spectrum.[35] If the (Supermax) prison is viewed as the pinnacle of a panopticon hierarchy, then the school is similarly imbued with panoptic impulses even if these are not permanent in their execution and effect.

Panopticonism can and should be interpreted as an ideal type, an accentuated construct of reality, or in other words, an abstracted representation of a phenomenon that can be employed as a standard for comparison. Importantly, Max Weber, asserts in relation to ideal types: 'In its conceptual purity, this mental construct cannot be found empirically anywhere in reality'.[36] Viewed in this way, panopticonism becomes an analytical tool and in abstraction retains its value as a heuristic for examining the contours of disciplinary mechanisms at work in the school. Foucault's panopticonism was 'a diagram of a mechanism of power', 'a figure of political technology',[37] and indeed we must treat it in this figurative sense. The panoptic schema is rhetorical, symbolic and emblematic of the power imbued in enclosed institutions.

An 'automatic functioning of power' was identifiable amongst some of the pupils in the case studies outlined in Chapter 2. For example, David, a pupil at City Comprehensive remarked: 'It [CCTV] encourages you to behave because you don't know if you are being watched through the camera or not. You're probably not, but you never know.' However, David intimates that the automatic functioning of power harnessed by the privilege of vision does not act equally upon all pupils. Some come to realise that rarely are displays of nonconformity responded to and so some 'just don't care' about the presence of the CCTV cameras. The pupils were often of the view that in all probability they are unlikely to be the focus of attention.[38] This rational calculation is the antithesis of Foucault's panopticonism.

Furthermore, Foucault's disciplinary institutions are characterised by 'little resistance'[39] and as such panopticonism does not allow much room for agency. The watched are passive subjects in a system of power over which they have no control. The previous chapter illustrates how there are often multiple channels of resistance to specific school strategies of discipline and control. However, it should be borne in mind that the majority of pupils do conform overall to the rules of the school: the majority arrive on time, wear the correct school uniform, walk in file, learn the subjects that they are instructed to, rise at the sound of the bell, eat at the allotted time, adopt the accepted movements (for example, raising a hand to ask a question, walking on the left of the corridor or queuing patiently for lunch) and so on.

Paradigm lost? Breaking out of the Panopticon

Metaphoric panopticonism has reached such hyperbolic proportions in surveillance studies that some now describe it as 'overwhelming',[40] 'a cliché',[41] 'oppressive'[42] and even as a theoretical 'straightjacket'.[43] With at least some reference to the Panopticon becoming de rigueur in the study of surveillance practices, it is certainly understandable why some feel fatigued by its incessancy. It has been suggested that the Panopticon has, somewhat ironically, come to exert an overbearing influence on studies of surveillance as the dominant theoretical framework.

> [T]he panoptic model has become reified, directing scholarly attention to a select subset of attributes of surveillance. In so doing analysts have excluded or neglected a host of other key qualities and processes of surveillance that fall outside of the panoptic framework.[44]

Kevin Haggerty implores us to 'tear down the walls' and 'demolish the panopticon'.[45] Whilst it is impossible to ignore the Panopticon in discussions of surveillance, it is possible to move beyond it.

The school plays a central role in disciplining children and this in itself is evocative of panopticonism, but the institutions in which disciplinary power originates cannot be equated to the unbounded territories of urbanity as some scholars have attempted to argue. The proliferation of surveillance technologies and practices represents a visible and convincing

DOI: 10.1057/9781137308863

'swarming of disciplinary mechanisms' from 'enclosed institutions' to the urban landscape. As such, a number of theorists claim that electronic surveillance, and in particular CCTV, can be seen to represent the infiltration of panopticonism into non-institutional populations.[46] However, simply lifting the disciplinary model of power from its institutional setting and attempting to directly apply it to the fluidity of society limits its applicability. After all, Panopticonism represents much more than an ability to observe those within a field of vision. Inherent in the panopticon is a disciplining and normalising function that transcends body and soul:

> The power of the panopticon is not just embodied in its ability to subject all to a surveillance gaze, but in the ability to link observation to a named subject through an individualized record, which can then be used for the purposes of identification, bureaucratic codification, and eventual classification. This secondary element, which in the Foucauldian sense transforms mere surveillance into discipline, is largely absent in the routine operation of CCTV systems.[47]

It has been argued that we must now talk of 'superpanopticonism',[48] 'hyper-panopticism'[49] or 'post-panopticonism'[50] but perhaps these are themselves exaggerated abstractions designed to emphasise the extremity of surveillance rather than a literal reading of current structures of power and their operation.

Although the school continues to propagate a culture of conformity and submission, it is important to recognise that the society in which it is now located is radically different to the modernist era. It has been opened up to new discourses of control and subjection beyond the correspondence between education and labour. Penal pedagogies and corporate priorities have increasingly flowed into schools as successive waves of neoliberalism have come to shape the political landscape.

Neoliberalism, urban renaissance and surveillance

A constellation of forces underpin the emergence of Surveillance Schools, but the shift to neoliberal post-industrialism permeates many causal vectors. This section posits two broad thematic trends that can be understood as symptomatic of the neoliberal regimen and are contributing to the ascension of Surveillance Schools: the criminalisation of schools and the corporatisation of schools.

DOI: 10.1057/9781137308863

The essence of neoliberalism is characterised by the retraction of non-market social entitlement, deregulation of the economy, and the privatising of state functions, institutions and public space. In *A Brief History of Neoliberalism* David Harvey[51] writes:

> neoliberalism is in the first instance a theory of political economic practices that proposes that human well-being can best be advanced by liberating individual entrepreneurial freedoms and skills within an institutional framework characterized by strong private property rights, free markets and free trade. The role of the state is to create and preserve an institutional framework appropriate to such practices.

Neoliberalism should not be viewed as a hegemonic economic model, but rather as 'an articulation of state, market, and citizenship that harnesses the first to impose the stamp of the second onto the third'.[52] The principles of the market have reached far beyond the realm of the economic, 'influencing and organising every domain of human activity and interaction'.[53] Converse to the utopian tropes of theoretical neoliberalism, in practice it has a polarising proclivity – liberalism for the upper classes and a 'punitive paternalism'[54] for the poor.

In 1977 Paul Willis described a classic exposition of cultural and social reproduction in the seminal text 'Learning to Labour'.[55] He argued that far from facilitating meritocratic social mobility, the school sorted pupils along the line of class thus explaining how 'working class lads get working class jobs'. More than three decades after Willis' ethnographic research, not surprisingly, the labour market and the society in which schools are situated have radically changed. Post-industrialism has brought with it a radical shift in the structure of the labour force. Jobs are now fragmented, uncertain, 'flexible' and mobile in a climate characterised by precarity due to 'outsourcing, just-in-time production, decentralization, computerized automation and temporary employment'.[56] Globalisation has charted the ascendancy of financial and speculative capital prioritising neoliberal ideals and displacing welfarism. Stratification is exemplified in the post-industrial labour economy which dichotomises between a mobile elite professional class, and an insecure, uncertain service-based labour force, or in other words a 'global "underclass" of unemployed/surplus workers'.[57] The rise of speculative and immaterial forms of labour in lieu of industrial labour furthers the 'birfurcation of national economies'[58] into categories of rich and poor, the latter of whom are subjected to enhanced social control and surveillance.

DOI: 10.1057/9781137308863

Clearly there are lessons to be taught in the post-industrial school beyond the correspondence between discipline and the workplace. As the neoliberal proclivity 'seeks to bring all human action into the domain of the market'[59] students are recast as 'human capital'[60] and the Surveillance School endeavours to sort them into two ideal types: complacent 'worker-consumers' and 'market rejects'. It is the latter which require increased surveillance, containment and control. The Surveillance School reflects the 'emergence of a state that is radically moving from a politics of social investment to a politics of containment'.[61]

Disinvestment in traditional public space and recreation has resulted in a shift towards corporately defined redevelopment priorities that utilise surveillance practices to safeguard the new 'bourgeois playgrounds'.[62] Dilapidated inner city communities are painted over with a 'triumphal gloss' to create new 'urban villages'.[63] Reconfigured economic and spatial arrangements in urbanised locales results in 'social sorting' supported by surveillance technologies to fulfil the 'exclusionary urge' characteristic of the regenerated urban environment which seeks to limit access only to those with the means and motivation to consume.[64] This can be most vividly demonstrated in the way that public city space, often regarded as democratic and open, has become increasingly regulated. In the urban renaissance, 'surveillance is used to 'sanitize' urban space, to exclude 'the Other' in a literal as well as metaphorical sense'.[65] In the neoliberal city, surveillance is appropriated as a tool to cleanse the market of 'undesirables' and make way for the 'legitimate consumer'.[66] Taken to its ultimate conclusion, the neoliberal urban renaissance creates 'fortress cities'; 'brutally divided between 'fortified cells' of affluent society where the police battle the criminalised poor.[67] The restructuring of the city results in 'economically and culturally disparate groups co-existing in relative close proximity'[68] which calls forth new structures of surveillance and social control to segregate polarised groups and placate tensions and nerves. The governance of urban space is bound up in a broader and more substantive 'social ordering' project with exclusionary consequences:

> Watching others has become systematic, embedded in a system that classifies according to certain pre-set criteria, and sorts into categories of risk or opportunity. These categories relate in turn to suspicion or solicitation – and many others in between – depending on the purposes for which surveillance is done.[69]

DOI: 10.1057/9781137308863

Surveillance is inherently exclusionary and seeks to sort individuals into discrete categorisations based on abstracted actuarial notions of risk. Young people are sorted along the lines of class, gender and race perpetuating and reinforcing social inequalities. Social exclusion, in which surveillance such as CCTV is complicit, attempts to render invisible unequal relations both on the streets and in political debate by sweeping away the mounting human debris of an advanced capitalist consumer society.

Examples of urban renaissance underpinned by networks of surveillance are numerous. Roy Coleman identified that shortly after Liverpool was awarded the title European Capital of Culture 2008, a drive to 'reclaim' city streets ensued.[70] This included passing a bylaw to curtail a number of different activities that could undermine the legitimacy of the neoliberal city such as grassroots and spontaneous street protests and skateboarding. The bylaws served to enhance and extend the segregative and exclusionary effects of neoliberal urbanity.[71] A similar process took place in Los Angeles, the 'City of Quartz', in its renaissance phase in which the outcast, discredited and incapacitated consumer was physically excluded from entire areas and communities, crystallised by the onset of private gated communities patrolled by armed guards. Those whose lifestyles are viewed as incompatible or a threat to consumption activities have their behaviour magnified through a 'demonological lens'[72] and so we begin to see bylaws that criminalise homelessness and the consumption of alcohol in public. Those who do not perform under the consumer-citizen rubric increasingly find themselves the focus of social control mechanisms and swathed in a conflux of surveillance practices to mitigate the risk they pose.

The net of deviance is being cast ever wider as the inherent inequalities of advanced capitalism require greater supervision to maintain order. Increasing control limits opportunities for individuals to contravene rules, and enables swift and harsh intervention should they do so. In the neoliberal regime 'social order is a matter not of shared values but of smart arrangements that minimise the opportunities for disruption and deviance'.[73] The growing number of behaviours that are classified as criminal, or the seeming mushrooming of some crimes over others, is not the result of a degenerate society brought about by teenage mothers, asylum seekers, terrorists, immigrants, the unemployed, uneducated, or an unruly and degenerate youth, but is rather a 'natural accompaniment', even embodiment, of the consumerist society.[74] An inescapable by-product of the neoliberal consumer society is the accompanying number of prisons literally bursting at the seams.

Post-industrial penality

Bolstering the increasingly punitive climate is a resurgence of the penal arm of the state signalling steps towards mass incarceration premised on judicial net-widening, up-tariffing and what Stanley Cohen evocatively terms 'mesh-thinning'.[75] More and more behaviours are being classified as criminal and subject to harsher penalties. In February 2008 the total prison population in England and Wales surpassed the useable operational capacity of the prison estate for the first time in history. More people per capita in England and Wales are imprisoned than ever before and future estimates do not make palatable reading. By the end of June 2017, it is projected to be as high as 94,800,[76] which, if attained, would involve the building of new prison facilities to house a new wave of incarcerated individuals. However, if England and Wales are pursuing a mass incarceration agenda, the United States has already undergone 'hyperincarceration'.[77] The number of people in custody in England and Wales is a drop in the ocean compared to the United States. Despite recent reductions, the US continues to be both the world's largest incarcerated population and hold the highest per capita incarceration rate. The Bureau of Justice Statistics estimates 6.98 million people were under some form of adult correctional supervision in the US in December 2011, the equivalent to approximately one in every 34 US adults (or about 2.9 percent of the adult population) being in prison, jail or on probation or parole.[78] Indeed 'the prison has made a spectacular comeback ... over the past three decades'.[79]

The 'imprisonment binge'[80] is not peculiar to England and Wales and the United States, but rather symptomatic of a global expansion of the penal state:

> With precious few exceptions ... incarceration has surged in all the postindustrial societies of the West, ballooned in the post-authoritarian nations of Latin America, and exploded in the nation-states issued from the collapse of the Soviet bloc as they transitioned from command to market economy. The carceral stock has not only risen rapidly ... along with the precarization of work and the retrenchment of welfare; it is everywhere composed disproportionately of the urban poor, ethnic and national outsiders, the homeless and the derelict mentally ill, and assorted rejects from the labor market.

DOI: 10.1057/9781137308863

It is widely accepted that the growing prison population is not being driven by increases in crime but by harsher sentencing in an increasingly punitive climate (if custody is viewed as the penological zenith). A new wave of technologically enhanced policing, surveillance and enforcement with an emphasis on punishment has been interpreted by some as evidence of a forever widening net of penal control. Garnered by the rhetoric of dangerous and risky schools, this 'penal swarming'[81] extends into schools through the outsourcing of discipline to the police and the emergence of a culture that increasingly resembles that of the prison.[82] Schools are increasingly being 'subsumed by a rapacious predilection for prison propagation'.[83]

Increasingly schools are becoming 'spaces in which students experience first-hand what it is to be monitored, feared, contained, and harassed and in the name of safety and protection'.[84] The school acculturates young people to this new reality by exposing them to penal pedagogies and a surveillance curriculum that habituates intense scrutiny from an early age. In the Surveillance School pupils are taught to expect law enforcement as a part of everyday life.[85] This moves the debate about Surveillance Schools beyond the parameters of crime control, even in its broadest sense, towards a more complex understanding of their purpose, application and consequence.

Risk, safety and penal pedagogies in the Surveillance School

Just as terrorism provided a fertile ground on which surveillance could propagate in the metropolis, moral panic about endemic school violence has provided a veneer of acceptability, even reverence, for the increased surveillance of pupils and policing of their environment. Crime has become the 'axis' around which the form and substance of schools has been recast and an incarceration mentality has infiltrated school discipline.[86]

Surveillance Schools are shrouded in discourses of 'safety' similar to the policies that underpin them. For example, in the US, policies are couched in emotive and compassionate terminology that conjures imagery of inclusive and supportive assistance, such as the Safe Schools Act 1994 and the No Child Left Behind (NCLB) Act in 2001. These policies in actual fact serve as vehicles to further the surveillance–security

DOI: 10.1057/9781137308863

agenda and activate pernicious policies and practices within the school environment such as zero tolerance approaches to school discipline. Sometimes, 'protection and oppression can be hard to tell apart'.[87] Surveillance in this respect can be seen as part of a 'governing process that constructs and amplifies particular risk messages'[88] and distils a climate of fear in urban schools.

Fear is amplified by the media, which enables the transformation of random acts of violence into a continuous narrative depicting a universal and imminent threat to (and from) pupils. Idiosyncratic events, such as school shootings, are transmitted to a global audience as further evidence of school campuses becoming increasingly dangerous territories. The violence that takes place in relatively few schools is then utilised as evidence of crime and risk in all schools. As the perceived problem acquires momentum, policy makers and law enforcers are required to act swiftly so as not to be held accountable for any future incidents. A self-perpetuating symbiotic spiral ensues as schools clamour for ever more sophisticated surveillance technologies to 'safeguard' pupils. For example, the supposed 'knife crime epidemic' in UK schools clearly underpinned the powers granted to school staff to search pupils without consent, and screen all pupils coming on site. Similarly, the justification for facial recognition technology has been rooted in the need to prevent 'another Dunblane'.

School shootings are often the 'go to' example used to present the 'need' for surveillance. For example, in reference to the Columbine High School massacre, Torin Monahan asserts that despite the failings of CCTV to prevent the atrocity 'the terrifying shooting has become a key reference point in justifying increased surveillance and security systems in schools throughout the United States'.[89] Such incidents provide currency for the expansion of techno-security mechanisms that serve to boost the surveillance economy.[90]

Similarly, in the wake of the tragic shootings at the Sandy Hook Elementary School in Connecticut in December 2012, school security was again under scrutiny. Despite state-led reviews of gun control legislation, some US states are implementing laws to enable teachers and school staff to carry guns.[91] With the United States' first Armed Teacher Training Program having commenced in April 2013 with teachers being taught gun skills and tactical manoeuvres it is now just a matter of time until a misunderstanding results in an innocent person being hurt. Such knee-jerk reactions do not tackle the underlying issues of school violence,

DOI: 10.1057/9781137308863

or address the potential for the seriousness of incidents to be exacerbated. In the wake of the Columbine High School shootings, Stephen Muzzati lamented that opportunities to discuss the 'culture of fear and violence, alienation, hyper-masculinity, frustration, marginalisation' and a host of other salient factors were missed.[92] The focus is always on target hardening and fortification with little consideration as to whether these strategies actually prevent future attacks from happening. The Sandy Hook Elementary School had a strong security procedure in place and was fitted with CCTV, however, it is reported that the perpetrator used a firearm to shoot out a window through which he then entered the building.

The 'if it saves just one child' mantra has become the policy equivalent of a full stop. A process of fear cultivation recasts expensive and disproportionate responses as 'necessary interventions, worth any cost, inconvenience, or more profound alteration of educational environments'[93] if it reduces the level of risk. But what if these 'rituals against fear' actually do very little to safeguard young people? Arguably, a false sense of security is worse than having none at all.

It is worth pointing out at this juncture that despite high profile violent events schools are in fact one of the safest places for children. In the US, statistically 'students are safer at school than they are in their own communities, in cars and even in their own homes'.[94] In terms of risk, it is poverty that continues to bring the greatest harm to children in developed nations. A study by the Johns Hopkins Center found that the 'lack of adequate health care may have contributed to the deaths of some 17,000 US children over the past two decades'.[95] In US schools 'not only is crime in schools down, but students also report feeling safer in schools today than they did a decade ago'.[96] There is ample evidence that the emergence of Surveillance Schools is not premised on a rise in crime:

> Although violence and disruption are unacceptable in schools and are hence key concerns that must be continually addressed in education, the evidence does not support an assumption that violence in schools is out-of-control. Serious and deadly violence remain a relatively small proportion of school disruptions, and the data have consistently indicated that school violence and disruption have remained stable, or even decreased somewhat, since approximately 1985.[97]

Furthermore, research has illustrated that over-reliance on technological equipment often manifests itself to the detriment of the school. In my own research at three schools in the North of England, I found that a 'diffusion

DOI: 10.1057/9781137308863

of responsibility' had materialised following the installation of CCTV whereby teachers and staff no longer intervened in incidents because they presumed the 'camera will capture it'. The quest for greater security can at times result in the 'simultaneous production of insecurity'.[98] As I have argued elsewhere about the exponential growth of CCTV,[99] surveillance technologies in schools are often welcomed as a visual representation that *something* is being done. It enables policy makers and education personnel to demonstrate that they are taking child safety seriously and are willing to invest in the latest techno-security to augment it. The discourse of safety is of course also politically expedient since it deflects attention away from the cultural, economic and political roots of school violence and provides a campaign vehicle with plenty of mileage for politicians. School violence quickly overshadows issues of persistent class and racial inequality.

The rhetoric of dangerous and risky schools not only drives a process in which fear and risk are drawn upon to rationalise and justify punitive sanctions, but assists in extending the reach of the penal state apparatus into schools (for example, in the guise of onsite police officers, metal detectors and sniffer dogs as outlined in Chapter 2). Behaviour in school is increasingly viewed through a criminal lens that requires discipline to be outsourced to external expert providers who, it is claimed in neoliberal philosophy, can provide a more efficient, effective and specialised service. Policies borrowed from criminal justice, surveillance equipment associated with criminality and onsite police officers bring with them new discourses and narratives that wrap around more traditional pedagogic knowledge and circulate in the school environment shaping the way young people view the school and society.

The market and the society in which schools are situated have radically changed, calling forth new lessons beyond the correspondence between discipline and the workplace. As the pipeline from school to the workplace has become fractured, the role of the educational institution has changed. The school-to-workplace pipeline has bottle-necked as neoliberalism desiccates jobs. The school production line is no longer routed only to the industrial workplace, but rather prisons have become the relief valve, filtering off the already marginalised poor into carceral warehouses. Post-industrialism has brought with it:

> the need for the management and control of a population that has become economically superfluous – particularly in urban deindustrialized zones... Urban public schools have become increasingly connected... to the criminal justice system and to the production of mass incarceration.[100]

There has been a radical shift from pupils 'learning to labour' to a school in which a proportion, especially in the United States are 'learning to do time'.[101]

Big Brother, Big Business: the corporatisation of schools

In August 2005, 11-year-old Rory Blackhall disappeared after his mother left him close to his Primary School in Livingston, Scotland. He was missing for three days before his body was found in the woods less than a mile away. He had been asphyxiated. Seizing the opportunity for marketing following this tragic event, just days after the boy had been found murdered Anteon UK Ltd contacted 340 local authorities to promote its 'VeriCool' school registration software with the following email:[102]

> *Dear Sir or Madam,*
>
> *Like everyone else, we were shocked and saddened by the apparent murder of the young schoolboy in West Lothian. We believe that we can help reduce the possibility of such future tragedies and so wish to bring to your attention our new anti-truancy and first day contact system that is already in use by some schools in the UK.*
>
> *I appreciate that you would not normally spotlight one type of technology from an unsolicited email, but our parent company, Anteon UK has worked with government departments on a number of occasions, and I genuinely feel that we are better placed than many to assist you in your target of reducing truancy and improving school children's security.*

The Advertising Standards Authority (ASA) described the marketing campaign as 'offensive and distressing', and furthermore deemed that Anteon UK Ltd had not been able to demonstrate that their product could have helped to prevent the tragedy.[103] What this email demonstrates, aside from the sinister side of marketing executives, is how lucrative the school surveillance economy is. A benevolent interest in school safety and the welfare of pupils is used as a Trojan horse to penetrate into the education market. Anteon UK Ltd are by no means alone in playing upon the fears and anxiety that surround schools in order to sell products. The number of companies that offer their services to safeguard schools is mushrooming and as a result schools are being reconfigured to follow the market logics of competition and corporate priorities. The cultivation of fear is played upon by manufacturers and suppliers of surveillance equipment to secure a market for their products. Big Brother is big business and there is always a need to find markets for new technological advances.

Manufacturers and suppliers can be seen to kick-start the surveillance arms race by donating equipment to run pilot programmes. For example, a school in Phoenix, Arizona, drew upon $350,000 worth of equipment donated by a local security company to implement biometric face scanners.[104] The scheme was not based on any evidence of efficacy, nor any demonstrated need for the equipment, highlighting that the consumption of surveillance equipment often 'has little to do with actual student safety and everything to do with business'.[105] There are many examples of seemingly benevolent gestures by vendors to assist the school in ensuring the safety of pupils and staff. Such loss leaders can serve to open up new markets for surveillance–security products and induce demand amongst schools. As some schools begin to develop strategies to counter the perceived risks, other schools quickly follow suit through fear of being regarded as negligent of their responsibilities *in loco parentis* if they didn't. This symbiotic process results in the perception that all schools 'need' ever more sophisticated technologies to 'safeguard' pupils. Once these systems are viewed as necessary, then any cost, whether financial or social, becomes worth the trade. It is an ingenious strategy to turn limited public funds into private profits. Not only do schools represent new lucrative markets, but they also provide a convenient test bed for new products. School pupils become the 'guinea pigs' for the latest security devices, and what better group to choose than a population who have very little command over their rights, and can be easily coerced and disciplined to comply?

It is difficult to ascertain exact figures but it is clear that the Surveillance School economy is booming. A survey of the security industry in the US found that the education sector was the third fastest growing market with a 15 per cent annual increase in sales.[106] It's not known how many schools use surveillance technologies, but they contribute a rapidly growing share of the $20 billion US security industry. Garrett Metal Detectors, described by Ronnie Casella in *Selling Us the Fortress* as 'perhaps the most successful metal detector company in the world',[107] began

DOI: 10.1057/9781137308863

selling equipment to schools in the 1990s. Since the Columbine shooting, schools have become the company's largest clients accounting for a quarter of the company's business. In 2003, sales to schools surpassed sales to prisons and airports. Their Vice President of marketing stated: 'We never dreamed there would be a market in schools. Now it's our largest-selling segment of our business... and I project that our sales to schools will probably double in the next year'.[108] Similarly, a BulletBlocker backpack, enforced with a bulletproof shield, saw a 300 to 400 per cent increase in sales since the Sandy Hook Elementary School tragedy.

New products are coming on to the market to increase, quite literally, the transparency of pupils and their activities. For example, in 2001, SeeThru Backpacks company was launched providing a product to enable school personnel to monitor the contents of pupil's backpacks from a distance. Technologies such as CCTV, iris scanners and palm vein readers that are typically associated with law enforcement and the military are increasingly finding that supplying Surveillance Schools with equipment is an attractive economic niche worth billions of dollars. Big Brother is Big Business and there are countless vendors with safety to sell.

Privatising schools

The rhetoric of dangerous and risky schools is politically expedient in furthering the agenda to privatise schools in both the US and Britain. As Ronnie Casella notes, 'we hear the same old free market story: for various reasons public officials cannot provide safety, or the safety they provide is inferior to what the private sector can provide, so the private sector must step in and pick up the slack'.[109] Schools are being restructured by market reforms, but safety is not the only mechanism by which the privatisation debate is garnered by surveillance.

In England, standardised testing and school league tables have been used as leverage in a move towards privatisation. Schools labelled as 'failing' are forced to become Academies: state-maintained schools that are independently run. Academies are sponsored or acquired by private entities, usually large chain foundations set up by millionaires. For example, the Harris Federation, founded by Lord Harris, the chairman of the retail chain Carpetright (and incidentally one of the Conservative party's biggest donors), has acquired more than 20 schools.[110] There has been considerable controversy around academisation. Due to the autonomy they are afforded, they can refuse entry to students that they perceive as low performing, including students with special needs, whilst cherry picking those from higher-achieving backgrounds. Exclusion rates are typically high and there have been claims that students are excluded so as not to impact on league tables.

While at the time of writing, Academies cannot be run for profit, but they can contract out school management to private for-profit companies. Mary Bousted, the general secretary of the Association of Teachers and Lecturers (ATL), argues that the real impetus for academisation of English schools 'is about the private sector coming in and raking off profits and assets from schools, taking taxpayers' money away from state education'.[111] The academisation of schools in Britain is almost certainly a precursor to their privatisation under the Coalition government, adhering to the neoliberal enterprise: 'if markets do not exist (in areas such as land, water, education, health care, social security, or environmental pollution) then they must be created, by state action if necessary'.[112] We are rapidly moving towards a situation in which students are treated as customers, and corporations propagate new market-oriented pedagogies.

Law, regulation and enforcement

Globally the regulation of surveillance is diverse. However, it is universal that as new ways of observing, scrutinising, recording and analysing individuals and their data emerge with ever increasing frequency, legal systems worldwide are unable to keep apace. The dearth and inadequacy of legislation coupled with the relative powerlessness of young people to reject their disproportionate subjection permits ever more invasive surveillance practices to be introduced in schools.

Surveillance Schools are flourishing concurrent with the 'vanishing rights of students'.[113] Drawing upon the UK as an example of a country with a proliferation of Surveillance Schools, it is clear that a lack of legal regulation has contributed towards their profusion. The European Union Data Protection Directive 1995 requires that members protect citizens' 'right to privacy with respect to the processing of personal data'. EU member states have implemented this in different ways. In the UK, the Directive manifested itself in the Data Protection Act 1998 (DPA 1998). In previous papers I have applied DPA 1998 decree to the widespread introduction of AFIS[114] and CCTV[115] in schools and found that the various elements of statute are impractical or inappropriate

DOI: 10.1057/9781137308863

in the school setting. Head teachers are vested with the autonomy to implement any technology they desire, and they are not legally obliged to gain the consent of parents, or even inform them. The ill-defined and vague legislation presented in the DPA 1998 provides very little protection to the pupils as 'data subjects'. In addition, the ubiquity of CCTV and AFIS far surpasses the enforcement capabilities of the Information Commissioner's Office (ICO), tasked with ensuring compliance with the DPA 1998, and as such any contravention of the scant provisions of the Act is likely to go unidentified and under enforced.

In the US, constitutional interpretation has evolved in the courts, materialising in a number of laws and executive orders dealing specifically with the concept of data protection. However, two developments of significance, the Privacy Act 1974 and the Computer Matching and Privacy Act, are applicable only to personal information held by the federal government and do not have any authority over the collection and use of personal information held by other private and public sector entities, including schools. The American Civil Liberties Union asserts that schools using CCTV are 'unconstitutionally intruding on the legitimate expectation of privacy of students, faculty, staff, and visitors', and are in effect 'engaging in unreasonable search without a warrant and without probable cause or reasonable suspicion'.[116] Irrespective of this declaration, it is unlikely that CCTV in schools will be ruled as unconstitutional in the US courts. The main reason compounding the issue is that historically, under the Fourth Amendment, young people are denied liberty rights because these rights presuppose the capacity to make informed choices, and the limited cognitive and emotional capacities of young children do not provide for this, it has been argued.

Schools are increasingly finding themselves on the intersection between penal pedagogies and corporate concerns. As institutions that have discipline at their core they are privileged in their ability to shape the views, actions and roles of future generations. It is in this capacity that public education must not be modelled on the corporate world. The current trajectory is privatisation in which the values of the corporation will displace social values, and citizenship will be further aligned with consumerist ideals. Pupils will be assessed according to their ability to contribute as worker/consumers, or disposed to poverty or prison. The final chapter offers some alternatives to the current disciplinary regime characterising Surveillance Schools.

DOI: 10.1057/9781137308863

Notes

1 Durkheim, E. (1925) *Moral Education: A Study in the Theory and Application of the Sociology of Education*. Glencoe, IL: The Free Press.
2 For example; Bowles, S. and Gintis, H. (1976) *Schooling in Capitalist America: Educational Reform and the Contradictions of Economic Life*. New York: Basic Books.
3 Lyon, D., ed. (2011) *Theorizing Surveillance: The Panopticon and Beyond*, 2nd edn. Oxon: Routledge, p. 4.
4 Garland, D. (1997) '"Governmentality" and the problem of crime: Foucault, criminology, sociology', *Theoretical Criminology* 1 (2): 173–214, p. 173.
5 Dill, J.S. (2007) 'Durkheim and Dewey and the challenge of contemporary moral education', *Journal of Moral Education* 36 (2): 221–37.
6 Durkheim, E. (1922) *Education and Sociology*. Glencoe, IL: The Free Press, p. 123.
7 Ibid., p. 81.
8 Bowles, S. and Gintis, H. (1976); Tyack, D.B. (1974) *The One Best System: A History of American Urban Education*. Cambridge, MA: Harvard University Press.
9 Foucault, M. (1977) *Discipline and Punish: The Birth of a Prison*. London: Penguin Books, p. 172.
10 Durkheim, E. (1922).
11 Durkheim, E. (1925), p. 47.
12 Ibid., p. 49.
13 Bowles, S. and Gintis, H. (1976), p. 37.
14 Foucault, M. (1977), p. 211.
15 Landon (1895) cited in Docking, J.W. (1980) *Control and Discipline in Schools*. London: Harper and Row Ltd, p. 16.
16 Ibid., p. 25.
17 Foucault, M. (1977), pp. 215–216.
18 A term Foucault (1977: 235) borrowed from Baltard, L. (1829) *Architectonographie des prisons*.
19 Foucault, M. (1977), p. 215.
20 Ibid., p. 138.
21 Ibid., p. 167.
22 Ibid., p. 215.
23 Norris, C. (2003) 'From the personal to the digital: CCTV, the panopticon, and the technological mediation of suspicion and social control', in D. Lyon (ed.), *Surveillance as Social Sorting. Privacy, Risk and Digital Discrimination*. London/New York: Routledge, 249–81, p. 249.
24 Boyne, R. (2000) 'Post-panopticonism', *Economy and Society* 29 (2): 285–307.
25 Foucault, M. (1977), p. 205.

DOI: 10.1057/9781137308863

26 Brunon-Ernst, A. and Tusseau, G. (2012) 'Epilogue: The Panopticon as a contemporary icon?', in A. Brunon-Ernst (ed.), *Beyond Foucault: New Perspectives on Bentham's Panopticon.* Farnham: Ashgate Publishing Limited, 185–200, p. 187.

27 Foucault, M. (1977), p. 155.

28 Ibid., p. 201.

29 Ibid., p. 211.

30 For example, see; Azzarito, L. (2009) 'The panopticon of physical education: pretty, active and ideally white', *Psychical Education and Sport Pedagogy* 14 (1): 19–39; Blackford, H. (2004) 'Playground Panopticonism: ring-around-the-children, a pocketful of women', *Childhood* 11 (2): 227–49; Bushnell, M. (2003) 'Teachers in the schoolhouse panopticon: complicity and resistance', *Education and Urban Society* 35 (3): 251–72; Perryman, J. (2006) 'Panoptic performativity and school inspection regimes: disciplinary mechanisms and life under special measures', *Journal of Education Policy* 21 (2): 147–61.

31 Gallagher, M. (2010) 'Are schools Panoptic?', *Surveillance & Society* 7 (3/4): 262–72, p. 264.

32 Welland, T. (2001) 'Living in the "empire of the gaze": time, enclosure and surveillance in a theological college', *The Sociological Review* 49 (1): 117–35.

33 Gallagher, M. (2010).

34 Foucault, M. (1977), p. 201.

35 Lyon, D. (2007) 'Resisting surveillance', in S.P. Hier and J. Greenberg (eds), *The Surveillance Studies Reader.* Berkshire: Open University Press, 368–77, p. 5.

36 Weber, M. (1949/2011) *Methodology of the Social Sciences.* ed. E.A. Shils and H.A. Finch, New Brunswick: Transaction Publishers, p. 90.

37 Foucault, M. (1977), p. 205.

38 Gill, M. and Loveday, K. (2003), p. 83.

39 Foucault, M. (1977), p. 218.

40 Haggerty, K.D. (2011) 'Tear down the walls: on demolishing the panopticon', in D. Lyon (ed.), *Theorizing Surveillance: The Panopticon and Beyond.* Oxon: Routledge, p. 23.

41 Hope, A. (2013) 'Foucault, panopticonism and school surveillance research', in M. Murphy (ed.), *Social Theory and Education Research: Understanding Foucault, Habermas, Bourdieu and Derrida* London: Routledge, 35–51, p. 35.

42 Haggerty, K.D. (2011), p. 23.

43 Aas, K.F., Gundus, H. O. And Lomell, H. M. (2009) 'Introduction: Technologies of (in)security', in H.O.G. K. F. Aas, and H. M. Lomell (eds), *Technologies of InSecurity: The Surveillance of Everyday Life.* Oxon: Routledge-Cavendish, 1–18, p. 3.

44 Haggerty, K.D. (2011), p. 23.

45 Ibid.

DOI: 10.1057/9781137308863

46 For example, see Fyfe, N.R. and Bannister, J. (1996) 'City watching: closed circuit television surveillance in public spaces', *Area* 28 (1): 37–46.
47 Norris, C. (2003) 'From the personal to the digital: CCTV, the panopticon, and the technological mediation of suspicion and social control', p. 256.
48 Poster, M. (1990) *The Mode of Information: Poststructuralism and Social Context.* Chicago: The University of Chicago Press.
49 Lewis, T. (2003) 'The surveillance economy of post-Columbine schools', *Review of Education, Pedagogy, and Cultural Studies* 25 (4): 335–55, p. 342.
50 Boyne, R. (2000).
51 Harvey, D. (2005) *A Brief History of Neoliberalism*. Oxford: Oxford University Press, pp. 3–4.
52 Wacquant, L. (2012) 'Three steps to a historical anthropology of actually existing neoliberalism', *Caderno CRH* 25: 505–18, p. 505.
53 Giroux, H.A. (2011) *On Critical Pedagogy*. London: Continuum International Publishing Group, p. 89.
54 Wacquant, L. (2012).
55 Willis, P. (1977) *Learning to Labour: How Working Class Kids Get Working Class Jobs*. Farnborough: Saxon House.
56 Kupchik, A. and Monahan, T. (2006) 'The New American School: preparation for post-industrial discipline', *British Journal of Sociology of Education* 27 (5): 617–31, p. 620.
57 Means, A. (2011) 'Securing the corporate state: education, economism, and crisis in the age of Obama', in K.J. Saltman and D.A. Gabbard (eds), *Education as Enforcement: The Militarization and Corporatization of Schools*. New York: Routledge, p. 263.
58 Nolan, K. and Anyon, J. (2004) 'Learning to do time: willis' cultural reproduction model in an era of deindustrialization, globalization, and the mass incarceration of people of color', in N. Dolby, G. Dimitriadis, and P. Willis (eds), *Learning to Labor in New Times.* New York: Routledge, 114–29.
59 Harvey, D. (2005).
60 Apple, M.W. (1998) 'Knowledge, pedagogy, and the conservative alliance', *Studies in the Literary Imagination* 31 (1): 5–23, p. 6.
61 Giroux, H. (2001) *Public Spaces, Private Lives: Beyond the Culture of Cynicism.* Lanham: Lowman and Littlefield, pp. 36–37.
62 Parenti, C. (2000) *Lockdown America: Police and Prisons in the Age of Crisis*, New York: Verso.
63 Davis, M. (1998) *City of Quartz: Excavating the Future in Los Angeles*. London: Pimlico, p. 224.
64 Coleman, R. (2004) *Reclaiming the Streets: Surveillance, Social Control and the City*. Devon: Willan publishing.
65 Koskela, H. (2003) '"Cam Era" – the contemporary urban Panopticon', *Surveillance & Society* 1 (3): 292–313, p. 298.

DOI: 10.1057/9781137308863

66 Bauman, Z. (1997) *Postmodernity and its Discontents*. Cambridge: Polity press.
67 Davis, M. (1998), p. 224.
68 Kupchik, A. and Monahan, T. (2006), p. 620.
69 Lyon, D. (2007) 'Resisting Surveillance', p. 371.
70 Coleman, R. (2004).
71 Beckett, K. and Herbert, S. (2008) 'Dealing with disorder: social control in the post-industrial city', *Theoretical Criminology* 12 (5): 5–30.
72 Davis, M. (1998), p. 224.
73 Garland, D. (2001) *The Culture of Control: Crime and Social Order in Contemporary Society*. Chicago: University of Chicago Press, p. 183.
74 Bauman, Z. (1997), p. 40.
75 Cohen, S. (1985) *Visions of Social Control*. Cambridge: Polity Press.
76 Ministry of Justice (2011) *Prison Population Projections 2011–2017 England and Wales*.
77 Wacquant, L. (2001) 'Deadly symbiosis: when ghetto and prison meet and mesh', *Punishment & Society* 3 (1): 95–134.
78 Glaze, L.E. and Parks, E. (2012) 'Correctional populations in the United States, 2011', *U.S. Department of Justice Bulletin November 2012*. Bureau of Justice Statistics.
79 Wacquant, L. (2012), p. 9.
80 Davis, A.Y. (2005) *Abolition Democracy: Beyond Empire, Prisons, and Torture*. New York: Seven Stories Press, p. 41.
81 Simon, J. (2007) *Governing Through Crime: How the War on Crime Transformed American Democracy and Created a Culture of Fear*. Oxford: Oxford University Press, p. 225.
82 Giroux, H.A. (2011) 'Foreword: Governing through crime and the pedagogy of punishment', in K.J. Saltman and D.A. Gabbard (eds), *Education as Enforcement: The Militarization and Corporatization of Schools*. New York: Routledge, vii–xvi, p. xi.
83 Lewis, T. (2003), p. 335.
84 Weiss, J., (2008), Under the Radar: School Surveillance and Youth Resistance. The City University of New York. Unpublished thesis. p. iv.
85 Simon, J. (2007).
86 Ibid., p. 209.
87 Hardt, M. and Negri, A. (2000) *Empire*. Cambridge, MA: Harvard University Press, p. 106.
88 Coleman, R. (2004), p. 6.
89 Monahan, T. (2006) 'The surveillance curriculum: risk management and social control in the neoliberal school', p. 109.
90 Lewis, T. (2003).
91 'South Dakota to allow armed teachers in schools', *BBC News*, 8 March 2013.

92 Muzzati, S.L. (2004) 'Criminalising marginality and resistance: Marilyn Manson, Columbine and cultural criminology', in J. Ferrell, et al. (eds), *Cultural Criminology Unleashed.* London: Glasshouse Press, 143–54.

93 Monahan, T. (2006), p. 113.

94 American Civil Liberties Union, 2001 cited in ibid, pp. 109–110.

95 Cited in: Giroux, H.A. (2011), p. 91.

96 Kupchik, A. (2010) *Homeroom Security: School Discipline in An Age of Fear.* New York: New York University Press, p. 15.

97 Skiba, R., Reynolds, C.L.R., Graham, S., Sheras, P., Conoley, J.C., and Garcia-Vazquez, E. (2006) 'Are zero tolerance policies effective in the schools? An evidentiary review and recommendations', American Psychological Association Zero Tolerance Task Force, 3–4.

98 Aas, K.F., Gundus, H. O. And Lomell, H. M. (2009), p. 1.

99 Taylor, E. (2010) 'Evaluating CCTV: why the findings are inconsistent, inconclusive and ultimately irrelevant', *Crime Prevention and Community Safety* 12: 209–32.

100 Nolan, K. and Anyon, J. (2004), 133–34.

101 Ibid.

102 Brook, S., 'Murdered schoolboy ad banned', *The Guardian*, 16 November 2005.

103 Ibid.

104 Kupchik, A. and Monahan, T. (2006).

105 Lewis, T. (2003), p. 350.

106 Cited in Fuentes, A. (2013) *Lockdown High: When the Schoolhouse Becomes a Jailhouse.* London: Verso, p. 133.

107 Casella, R. (2006), p. 126.

108 Cited in Lewis, T. (2003), p. 336.

109 Casella, R. (2010) 'Safety or social control? The security fortification of schools in capitalist society', in T. Monahan and R.D. Torres (eds), *Schools Under Surveillance.* New Brunswick: Rutgers University Press, 73–86, p. 84.

110 Monbiot, G., 'With threats and bribes, Gove forces schools to accept his phoney "freedom"', *The Guardian*, 4 March 2013.

111 Cited in Vasagar, J. and Mansell, W., 'Private firm sniffs a profit in coalition schools policy', *The Guardian*, 20 June 2011.

112 Harvey, D. (2005), p. 2.

113 Beger, R.R. (2002).

114 Taylor, E. (2010) 'From finger-painting to fingerprinting: the use of biometric technology in schools', *Education Law Journal* 4: 276–88.

115 Taylor, E. (2010) 'UK schools, CCTV and the Data Protection Act 1998', *Journal of Education Policy* 26 (1): 1–15.

116 American Civil Liberties Union, Chicago, cited in Warnick, R. (2010), p. 320.

DOI: 10.1057/9781137308863

Conclusion

Abstract: *There is a need to rethink school discipline. Restorative Justice is offered as an effective and desirable philosophy that shifts focus away from punitive punishments towards a more inclusive approach to reparative discipline. Furthermore, schools must place surveillance studies firmly on the curriculum to ensure that young people are able to engage critically with its operation and understand how to interact with it in their everyday lives.*

Keywords: Restorative Justice, surveillance curriculum, habituation, school discipline

Taylor, Emmeline. *Surveillance Schools: Security, Discipline and Control in Contemporary Education.* Basingstoke: Palgrave Macmillan, 2013.
DOI: 10.1057/9781137308863.

A growing body of literature has demonstrated how surveillance in schools can undermine privacy, erode trust, have a 'chilling effect' on creativity and interaction, criminalise students, and in the most extreme cases facilitate a direct and expedited channel from the school to prison. The evidence that is emerging from empirical studies suggests that often the technologies do little to safeguard young people, do not represent financial savings or increased efficiencies, but yet the incessant use of omnipresent technological surveillance has the ability to displace the very building blocks of democratic society. There are concerns about the vanishing rights of children as the Surveillance School relinquishes young peoples' rights to personal freedom, anonymity and privacy. For some, young people are no longer considered the world's future, but as a threat to the present.[1] A cross-national analysis of the surveillance of young people found that the current generation are growing up with a profound mistrust of the state, a diminished belief in democracy, 'cynical views about a broad-based "common good,"' and 'little memory of or imagination for a public sphere for the public'.[2] Embedded in Surveillance Schools are structural forms of violence against youth that contribute towards their marginalisation, exclusion and disenfranchisement. Is it any surprise that young people are increasingly sceptical about democracy?

Jonathan Simon claims that in the early 1990s most US schools remained highly protective of students, avoiding sanctions such as suspension and expulsion where possible, knowing that it would hugely disadvantage their future prospects.[3] School discipline was viewed as distinct and separate from the criminal justice system and young people experiencing problems were offered assistance rather than being automatically viewed through a criminal lens. This time appears to have now passed. Young people are being squeezed into more confined and regulated spaces with less room in which they can be children. Innocence, naivety and even childish foolishness are not viewed as an essential part of growing up, but treated in the same way as deliberate acts of harm perpetrated by adults. In the Surveillance School, 'punishment and policing have come to at least compete with, if not replace, teaching as the dominant modes of socialization'.[4] Punitive pedagogies are displacing discourses of compassion, welfare and well-being. As neoliberalism has become 'hegemonic as a mode of discourse'[5] it has enabled corporate schemas to pervade schools, reclassifying citizens as consumers and aligning education with the needs of the post-industrial market.

DOI: 10.1057/9781137308863

There is a pressing need to rethink school discipline and the appropriateness of embedding surveillance technologies into pedagogical apparatus. The purpose of education should not be to train students to perform useful roles in the corporate order, but rather to produce human beings that are conscientious, aware and capable of contributing to a society based on democratic and inclusive principles. Below I outline a potential avenue for rethinking school discipline.

Rethinking school discipline

Overzealous disciplinary techniques such as zero tolerance approaches create schools for docility rather than spaces for creative experimentation and expression. Compliance must be borne out of compassion, understanding and a sense of community rather than through fear of the repercussions for misbehaviour. In today's rapidly changing world, young people need to be taught the skills required to navigate changing social expectations, complex interactions and transient roles. Young people inhabit a global world and one increasingly premised on uncertainty. Providing students with the tools to be able to respond to new situations in a rational, logical and appropriate manner is imperative. This process involves nurturing critical thought. It cannot be achieved through 'teaching to the test', scripted lessons or by refusing to engage with the issues that pupils raise about their surveillance. If future generations are to emerge from schools as well-rounded social participants, they need to be treated with respect in order to learn how to respect themselves and the others around them.

Restorative Justice

Restorative Justice has received growing attention as an alternative to traditional criminal justice meted out by state institutions such as the police or prisons. It can be conceived of as both a process and a value system. In the former sense, Restorative Justice is 'a process that brings together all stakeholders affected by some harm that has been done (offenders, their families, victims and their families, affected communities, state agencies such as the police and so on). These stakeholders meet in a circle to discuss how they have been affected by the harm and come to some agreement as to what should be done to right any wrongs suffered.'[6] In the latter sense, as a value system, Restorative Justice is focused on repairing harms, rather than responding to hurts inflicted with 'the hurt of punishment'.[7] It represents a paradigm shift from social control to social engagement.

Restorative Justice as an approach to school discipline has received increasing attention over the past decade. Education theorists have long recognised the need to have a structure of discipline in schools that is viewed as just, fair and proportionate to transgressive behaviour. Lisa Cameron and Margaret Thorsborne argue that a Restorative Justice philosophy in schools can offer such attributes to school discipline by:

> providing opportunities for insight and learning when behaviour is deemed unacceptable; providing opportunities for dialogue and reflection when behaviour threatens the social cohesion of the school community; and identifying issues of harm to relationships and how to 'make things right' through strengthening relationships.[8]

The process has been found to be effective in schools, particularly in regard to tackling problem youth behaviour such as bullying.[9] Two evaluation studies of Restorative Justice in 119 schools across Queensland, Australia found that there was a high level of participant satisfaction with the process including, greater sense of justice, greater support for those affected and reduced reoffending.[10]

A Restorative Justice approach to school discipline can assist with cultivating trust, reducing conflicts, developing citizenship skills, promoting self-esteem and restoring the belief in democratic values. Above all else, it cultivates hope.[11] In *Just Schools*,[12] Belinda Hopkins and Guy Master have developed a comprehensive overview of how to introduce a Restorative Justice approach in the school context as a means of rejecting the authoritarian, punitive punishment orientated climate that has engulfed education.

Surveillance on the curriculum

In a global media age, in which data profiles are increasingly used to verify and authenticate identity, young people need to be taught the importance of their personal data and the potential repercussions of sharing it. However, the Surveillance School teaches young people to be casual about biometrics and data when they are applied to mundane activities. At the time of writing, Britain's first Youth Police and Crime Commissioner had just been appointed. The remit of the role was to attempt to bridge gaps between young people and the police, and represent their views in

DOI: 10.1057/9781137308863

key policy areas. It was a positive development at a time that relations between youth and criminal justice were at a low. It was an approach that could set a precedent across Britain on how to engage young people and provide them with a voice to influence and shape policy that impacted upon them. However, the appointed candidate, not having even started her first day on the job received a barrage of media attention. The reason? Her Twitter account had been accessed by journalists and a number of Tweets that she had posted between the ages of 14 and 16 were circulated across media outlets and described as homophobic, racist and violent. Indeed her posts did reflect some distasteful, prejudice and offensive remarks which are not excusable, but what the incident really brought to the fore was the lack of understanding regarding social media forums amongst its millions of users. She stated: 'I have fallen into the trap of behaving with bravado on social-networking sites. I hope this may stand as a learning experience for many other young people'. There is a desperate need to include surveillance on the curriculum and begin to educate young people about their responsibilities when engaging with such mediums. Again, responsibility does not come through punishment, but through educating for compassion, respect and empathy.

Habituation and normalisation

Back in 2006 when I was conducting interviews in schools, the pupils told me that they felt criminalised by the use of CCTV in their schools. They were disappointed with the lack of trust that they were given and felt that they were always under suspicion. Aside from that they were able to articulate with great insight and clarity how they valued privacy and how this was being taken away from them. Similar findings have been reported in other studies on youth perceptions of surveillance.[13] Fast forward to 2013 and I wonder whether the research would produce the same findings. Do pupils still feel criminalised by the presence of CCTV in schools? Do they still perceive they are receiving the same treatment as offenders when they are required to present a fingerprint for a library book? Do they believe that uniformed police and probation officers are symptomatic of a mistrust of them? In all honesty 'probably not' is the answer. I have informally surveyed students whom I teach at the Australian National University, and spoken with students at many high schools, and all too often the view is a 'so what?' approach to the

DOI: 10.1057/9781137308863

intensifying surveillance in their schools. How can this be interpreted? Have young people become habituated to surveillance as I argued they would be? Empirically it is difficult to determine.

'We quickly grow used to the way things are', asserts David Garland in *The Culture of Control*.[14] Increased monitoring and scrutiny arguably have a normalising effect on subjectivity. Without adequate regulation critical opportunities to question, redefine and reject the infiltration of surveillance technologies into the school environment are being missed. Andre Bacard laments that if one wanted to develop a surveillance society they should 'start by creating dossiers on kindergarten children so the next generation couldn't comprehend a world without surveillance'.[15] It sounds almost twee in the context of the contemporary Surveillance School equipped with CCTV, fingerprinting and radio frequency identification. Examining the normalisation and appropriation of surveillance by those that are subjected to it is paramount in assessing whether surveillance practices in schools are part of an offensive to subjugate the fears of society about 'surveillance creep' by teaching the next generation to accept increasing scrutiny. Society is on a trajectory towards accepting surveillance in schools, with the little iotas of resistance being eroded all of the time.

Perhaps surveillance in schools can be interpreted as a microcosmic institutional version of Zygmunt Bauman's 'There Is No Alternative' (TINA) thesis written small. Surveillance has become a *fait accompli* and before long schools that are not supplemented with automated surveillance technologies, fortified with CCTV cameras or patrolled by dedicated police officers appear naïve, quaint and old-fashioned. In 2010, following a number of exposés in England, there was considerable interest and media furore around the fingerprinting of school children without parental consent. However, just a few years later, the concern has dwindled and interest has depleted; it has become old news.

Upcoming generations will emerge from Surveillance Schools desensitised to, and expectant of, intense scrutiny and objectification. They will have no experience or comprehension of a world without invasive surveillance for even the most routine of activities such as purchasing a sandwich at lunchtime or borrowing a book from the library. The ramifications are huge. We are stood on the verge of what could be one of the most colossal shifts in societal structure and organisation; technological surveillance has the potential to redefine interpersonal relationships and the essence of social interaction and nowhere is the surveillance revolution more crystallised than in schools.

DOI: 10.1057/9781137308863

In the Introduction I cautioned against falling into the dystopian trap. Whilst this book detailed the potential negative impacts of school surveillance and their unsavoury ideological underpinnings, it should not be read as a plea to reject all surveillance technologies. There is a need for critical engagement with the positives as well as the negative implications; 'rather than demonizing and rejecting new technologies out of hand, we should criticise their misuse but also see how they can be used constructively for positive ends'.[16] Almost everyone reading this book will have had some experience of elementary or high school. What is remarkable is that the current generation are implementing surveillance technologies and practices that when we were students would have been regarded as invasive and unjustified. It's incredible just how quickly surveillance becomes a customary and indispensible practice when applied to the 'other', less powerful groups.

Surveillance Schools are already equipped with CCTV, fingerprinting, and radio frequency identification as well as a raft of other surveillance technologies and practices; the process of habituation is well under way and the opportunity for critical discussion is fast disappearing. We are moving past critical questions about the desirability of surveillance technologies displacing face-to-face interaction with teachers and school staff, and moving beyond debates about whether this is the future we want for schools. Those raising concerns or evening asking questions are deemed as old-fashioned, Luddites even, that are fearful of change and progress. Why else would they question developments that are intended to safeguard pupils, create efficiencies and save resources? Unless they have something to hide of course.

The future of Surveillance Schools

I started this book with a fictitious scenario in order to convey the level of surveillance currently being used in schools, but the current picture is still fractured and disjointed. Surveillance Schools have gathered momentum over the past decade but they are still very much in their infancy. Only time will tell how Surveillance Schools might develop in the not-too-distant future as the use of surveillance practices become normalised, as surveillance technologies get cheaper and more pervasive, and as neoliberalism cements the link between corporatisation and education. Schools as microcosms of society can permit us a prophetic

DOI: 10.1057/9781137308863

glimpse into the future, and as such it is a crucial time for scholars to study the phenomena of the 'Surveillance School'.

Notes

1 Alain Bertho cited in Durand, J.-M., 'For Youth: A Disciplinary Discourse Only', *TruthOut*, 15 November 2009. Available at: http://archive.truthout.org/11190911 (accessed 23 April 2013).
2 Ruck, M., Harris, A., Fine, M., and Freudenberg, N. (2008) 'Youth experiences of surveillance: a cross-national analysis', in M. Flynn and D.C. Brotherton (eds), *Globalizing the Streets: Cross-Cultural Perspectives on Youth, Social Control, and Empowerment*. New York: Columbia University Press, 15–30, pp. 15–16.
3 Simon, J. (2007) *Governing Through Crime: How the War on Crime Transformed American Democracy and Created a Culture of Fear*. Oxford: Oxford University Press.
4 Ibid., p. 210.
5 Harvey, D. (2005) *A Brief History of Neoliberalism*. Oxford: Oxford University Press, p. 4.
6 Braithwaite, J. and Strang, H. (2001) 'Introduction: Restorative Justice and civil society', in H. Strang and J. Braithwaite (eds), *Restorative Justice and Civil Society*. Cambridge: Cambridge University Press, 1–13, p. 1.
7 Ibid.
8 Cameron, L. and Thorsborne, M. (2001) 'Restorative justice and school discipline: Mutually exclusive?', in H. Stang and J. Braithwaite (eds), *Restorative Justice and Civil Society*. Cambridge: Cambridge University Press. 180–94, p. 180.
9 Ibid.
10 Ibid.
11 Braithwaite, J. and Strang, H. (2001).
12 Hopkins, B. and Master, G. (2004) *Just Schools: A Whole School Approach to Restorative Justice*. Gateshead: Athenaeum Press.
13 For example, see Mccahill, M. and Finn, R. (2010) 'The social impact of surveillance in three uk schools: "angels", "devils" and "teen mums"', *Surveillance & Society* 7 (3/4): 273–89.
14 Garland, D. (1997) "Governmentality' and the problem of crime: Foucault, criminology, sociology', *Theoretical Criminology* 1 (2): 173–214, p. 173.
15 Bacard, A. (1995) *The Computer Privacy Handbook: A Practical Guide to Email Encryption, Data Protection and PGP Privacy Software*. New York: Peachpit Press, p. 173.
16 Kellner, D. (2000) 'Multiple Literacies and Critical Pedagogies', in P. Trifonas (ed.), *Revolutionary Pedagogies*. New York: Routledge, p. 215.

DOI: 10.1057/9781137308863

Bibliography

Aas, K.F., Gundus, H. O. And Lomell, H. M. (2009) 'Introduction: Technologies of (in)security', in H.O.G. K. F. Aas, and H. M. Lomell (eds), *Technologies of InSecurity: The Surveillance of Everyday Life.* Oxon: Routledge-Cavendish.

American Civil Liberties Union (2002) *ACLU drug testing cases across the nation. Factsheet 15.* Available at: www.aclu.org/DrugPolicy/DrugPolicy.cfm?ID=11001&c=79 (accessed 18 March 2013).

Apple, M.W. (1982) *Education and Power.* London: Routledge.

Apple, M.W. (1998) 'Knowledge, pedagogy, and the conservative alliance', *Studies in the Literary Imagination* 31 (1): 5–23.

Armitage, R., Smyth, G., and Pease, K. (1999) 'Burnley CCTV evaluation', in K. Painter and N. Tilley (eds), *Surveillance of Public Space: CCTV, Street Lighting and Crime Prevention.* New York: Criminal Justice Press, 225–49.

Association of Teachers and Lecturers (2008) '*85% of teachers have CCTV in their schools and nearly 25% worry about hidden cameras*'. Available at: www.atl.org.uk/media-office/media-archive/CCTV-in-schools.asp (accessed 04 September 2013).

Azzarito, L. (2009) 'The panopticon of physical education: pretty, active and ideally white', *Psychical Education and Sport Pedagogy* 14 (1): 19–39.

Bacard, A. (1995) *The Computer Privacy Handbook: A Practical Guide to Email Encryption, Data Protection and PGP Privacy Software.* New York: Peachpit Press.

DOI: 10.1057/9781137308863

Ball, S.J. (1987) *The Micro-Politics of the School*, 2nd edn. London: Methuen & Co. Ltd.

Barkham, P. (14 May 2005) 'How a top can turn a teen into a hoodlum', *The Guardian*.

Barnard-Wills, D. (2012) 'E-safety education: young people, surveillance and responsibility', *Criminology and Criminal Justice* 12 (3): 239–55.

Barr, R. and Pease, K. (1990) 'Crime Placement, Displacement and Deflection', in M. Tonry and N. Morris (eds), *Crime and Justice: A Review of Research*. Chicago, IL: University of Chicago Press.

Bauman, Z. (1997) *Postmodernity and Its Discontents*. Cambridge: Polity press.

Bauman, Z. and Lyon, D. (2013) *Liquid Surveillance*. Cambridge: Polity Press.

Bawden cited in Garner, R., '"Hoodies, louts, scum": how media demonises teenagers', *The Independent*, 13 March 2009.

BBC News (15 September 2006) 'Parents feed pupils through gates'.

BBC News (11 April 2008) 'CCTV could be used in exams rooms'.

BBC News (4 March 2009) 'Teachers watched on CCTV cameras'.

BBC News (27 January 2009) 'School head defends toilets CCTV'.

BBC News (8 March 2013) 'South Dakota to allow armed teachers in schools'.

Beckett, K. and Herbert, S. (2008) 'Dealing with disorder: social control in the post-industrial city', *Theoretical Criminology* 12 (5): 5–30.

Beger, R.R. (2002) 'Expansion of police power in public schools and the vanishing rights of students', *Social Justice* 29: 119–30.

Bennett, C. and Raab, C. (2007) 'The privacy paradigm', in S.P. Hier and J. Greenberg (eds), *The Surveillance Studies Reader*. Berkshire: Open University Press.

BigBrotherWatch (2012) *Class of 1984: The Extent of CCTV in Secondary Schools and Academies*.

Blackford, H. (2004) 'Playground Panopticonism: ring-around-the-children, a pocketful of women', *Childhood* 11 (2): 227–49.

Bloss, W. (2007) 'Escalating U.S. police surveillance after 9/11: an examination of causes and effects', *Surveillance & Society* 4 (3): 208–28.

Bowles, S. and Gintis, H. (1976) *Schooling in Capitalist America: Educational Reform and the Contradictions of Economic Life*. New York: Basic Books.

Boyne, R. (2000) 'Post-panopticonism', *Economy and Society* 29 (2): 285–307.

DOI: 10.1057/9781137308863

boyd, D. and Marnick, A. (2011) 'Social steganography: privacy in networked publics', *Paper presented at ICA*, 28 May, Boston, MA.
Braithwaite, J. and Strang, H. (2001) 'Introduction: Restorative Justice and civil society', in H. Strang and J. Braithwaite (eds), *Restorative Justice and Civil Society*. Cambridge: Cambridge University Press.
Brunon-Ernst, A. and Tusseau, G. (2012) 'Epilogue: the Panopticon as a contemporary icon?', in A. Brunon-Ernst (ed.), *Beyond Foucault: New Perspectives on Bentham's Panopticon.* Farnham: Ashgate Publishing Limited, 185–200.
Bushnell, M. (2003) 'Teachers in the schoolhouse panopticon: complicity and resistance', *Education and Urban Society* 35 (3): 251–72.
Cameron, K. (2007) 'Just lie so you can sell your product', *Kim Cameron's Identity Weblog. Available at: www.identityblog.com* (accessed 03 September 13).
Casella, R. (2006) *Selling Us the Fortress*. Oxon: Routledge.
Casella, R. (2010) 'Safety or social control? The security fortification of schools in capitalist society', in T. Monahan and R.D. Torres (eds), *Schools Under Surveillance*. New Brunswick: Rutgers University Press.
CBBC (1 June 2007) 'Does CCTV make you feel safer?', *Newsround.* Available at: http://news.bbc.co.uk/cbbcnews/hi/newsid_6700000/newsid_6700500/6700513.stm (accessed 08 April 2013).
CBBC (25 July 2002) 'Should kids get fingerprinted?', *Newsround.* Available at: http://news.bbc.co.uk/cbbcnews/hi/chat/your_comments/newsid_2146000/2146389.stm (accessed 08 April 2013).
Chamberlin, T., 'Guns in schools', *The Sunday Mail*, 30 September 2012.
Clark, L., 'One in three secondary schools fingerprinting pupils as Big Brother regime sweeps education system', *Daily Mail Online*, 9 June 2010.
Cohen, S. (1985) *Visions of Social Control.* Cambridge: Polity Press.
Coleman, R. (2004) *Reclaiming the Streets: Surveillance, Social Control and the City*. Devon: Willan publishing.
Corrigan, P. (1979) *Schooling the Smash Street Kids*. London: Macmillan Press.
Crawford, A. (2000) 'Situational crime prevention, urban governance and trust relations', in A. Von Hirsch, D. Garland, and A. Wakefield (eds), *Ethical and Social Perspectives on Situational Crime Prevention.* Oxford: Hart Publishing.
Davies, S., 'Law that favours disorder', *The Guardian*, 21 September 2002.

DOI: 10.1057/9781137308863

Davis, A.Y. (2005) *Abolition Democracy: Beyond Empire, Prisons, and Torture*. New York: Seven Stories Press.

Davis, M. (1998) *City of Quartz: Excavating the Future in Los Angeles*. London: Pimlico.

Demarche, E., 'Pennsylvania girl, 5, suspended for threatening to shoot girl with pink toy gun that blows soapy bubbles', *FoxNews*, 19 January 2013.

Devine, J. (1996) *Maximum Security: The Culture of Violence in Inner-City Schools*. Chicago: University of Chicago Press.

Dill, J.S. (2007) 'Durkheim and Dewey and the challenge of contemporary moral education', *Journal of Moral Education* 36 (2): 221–37.

Docking, J.W. (1980) *Control and Discipline in Schools*. London: Harper and Row Ltd.

Dohrn, B. (2001) '"Look out kid/It's something you did": zero tolerance for children', in W. Ayers, B. Dohrn, and E. Ayers (eds), *Zero Tolerance: Resisting the Drive for Punishment in Our Schools*. New York: The New Press, 89–113.

Duel, M. (10 September 2012) 'Fury as school with second-highest expulsion rate in England builds "prison block" to house 12 troublemaker pupils', *The Daily Mail*.

Durkheim, E. (1922) *Education and Sociology*. Glencoe, IL: The Free Press

Durkheim, E. (1925) *Moral Education: A Study in the Theory and Application of the Sociology of Education*. Glencoe, IL: The Free Press.

Echo (2009) 'Teenage boys and the media', *www.womeninjournalism.co.uk* (accessed 09 April 2013).

Ericson, V. and Haggerty, K.D. (1997) *Policing the Risk Society*. Oxford: Oxford University Press.

European Union Directorate General for Research (1998) *An Appraisal For Technology of Political Control – Report*. Brussels: EUDGR.

Feeley, M. (2003) 'Crime, social order and the rise of neo-Conservative politics', *Theoretical Criminology* 7 (1): 111–30.

Foucault, M. (1977) *Discipline and Punish: The Birth of a Prison*. London: Penguin Books.

Foxnews.Com, 'Missouri high school reportedly to use hair samples for random drug tests', 31 January 2013.

Fried, C. (1971) *An Anatomy of Values: Problems of Personal and Social Choices*. Cambridge: Cambridge University Press.

DOI: 10.1057/9781137308863

Fuentes, A. (2013) *Lockdown High: When the Schoolhouse Becomes a Jailhouse*. London: Verso.

Furedi, F. (1997) *Culture of Fear: Risk Taking and the Morality of Low Expectation*. London: Cassell.

Fyfe, N.R. and Bannister, J. (1996) 'City watching: closed circuit television surveillance in public spaces', *Area* 28 (1): 37–46.

Gallagher, M. (2010) 'Are schools Panoptic?', *Surveillance & Society* 7 (3/4): 262–72.

Garland, D. (1997) '"Governmentality" and the problem of crime: Foucault, criminology, sociology', *Theoretical Criminology* 1 (2): 173–214.

Garland, D. (2001) *The Culture of Control: Crime and Social Order in Contemporary Society*. Chicago: University of Chicago Press.

Gerada, C. (2005) 'Random drug testing in schools', *British Journal of General Practice*, 55 (516): 499–501.

Gerstein, R. (1984) 'Intimacy and privacy', in F. Schoeman (ed.), *Philosophical Dimensions Privacy: An Anthology.* Cambridge: Cambridge University Press.

Giddens, A. (1985) *A Contemporary Critique of Historical Materialism: The Nation-state and Violence*. vol. 2, Cambridge: Polity.

Gill, M. and Loveday, K. (2003) 'What do offenders think about CCTV?', in M. Gill (ed.), *CCTV.* Leicester: Perpetuity Press, 81–92.

Giroux, H.A.(2001) *Public Spaces, Private Lives: Beyond the Culture of Cynicism*. Lanham: Rowman and Littlefield.

Giroux, H.A. (2003) 'Racial injustice and disposable youth in the age of zero tolerance', *International Journal of Qualitative Studies in Education*, 16 (4): 553–65.

Giroux, H.A. (2011) 'Foreword: Governing through crime and the pedagogy of punishment', in K.J. Saltman and D.A. Gabbard (eds), *Education as Enforcement: The Militarization and Corporatization of Schools.* New York: Routledge, vii–xvi.

Giroux, H.A. (2011) *On Critical Pedagogy*. London: Continuum International Publishing Group.

Glaze, L.E. and Parks, E. (2012) 'Correctional populations in the United States, 2011', *U.S. Department of Justice Bulletin November 2012.* Bureau of Justice Statistics.

Goffman, E. (1959) *The Presentation of the Self in Everyday Life*. London: Penguin.

DOI: 10.1057/9781137308863

Graham, S. (1998) 'Towards the fifth Utility? On the extension and Normalisation of Public CCTV', in J. M. C. Norris, and G. Armstrong (eds), *Surveillance, Closed Circuit Television and Social Control.* Aldershot: Ashgate.

Groombridge, N. (2002) 'Crime control or crime culture TV?', *Surveillance & Society* 1 (1): 30–46, p. 43.

Haggerty, K.D. and Ericson, R.V. (2000) 'The surveillant assemblage', *British Journal of Sociology*, 51 (4): 605–22.

Haggerty, K.D. (2011) 'Tear down the walls: on demolishing the panopticon', in D. Lyon (ed.), *Theorizing Surveillance: The Panopticon and Beyond.* Oxon: Routledge.

Hardt, M. and Negri, A. (2000) *Empire.* Cambridge, MA: Harvard University Press.

Harvey, D. (2005) *A Brief History of Neoliberalism.* Oxford: Oxford University Press.

Heidegger, M. (1977) *The Question Concerning Technology and Other Essays.* London: Harper and Row.

Hier, S.P. and Greenberg, J. (eds) (2007), *The Surveillance Studies Reader.* Berkshire: Open University Press.

Hinds, L. and Grabosky, P. (2008) 'Responsibilisation revisited: from concept to attribution in crimecontrol', *Security Journal* 23 (2): 95–113, p. 99.

Hirschfield, P.J. (2008) 'Preparing for prison? The criminalization of school discipline in the USA', *Theoretical Criminology* 12: 79–101.

Hope, A. (2005) 'Panopticonism, play and the resistance of surveillance: case studies of the observation of student Internet use in UK schools', *British Journal of Sociology of Education* 26 (3): 359–73.

Hope, A. (2008) 'Internet pollution discourses, exclusionary practices and the "culture of over-blocking" within UK schools', *Technology, Pedagogy and Education* 17 (2): 103–13.

Hope, A. (2010) 'Student resistance to the surveillance curriculum', *International Studies in Sociology of Education* 20 (4): 319–34.

Hope, A. (2013) 'Foucault, panopticonism and school surveillance research', in M. Murphy (ed.), *Social Theory and Education Research: understanding Foucault, Habermas, Bourdieu and Derrida.* London: Routledge.

Hopkins, B. and Master, G. (2004) *Just Schools: A Whole School Approach to Restorative Justice.* Gateshead: Athenaeum Press.

DOI: 10.1057/9781137308863

Indymedia.org.uk (2006) 'Student fingerprinting – not so Vericool'. Available at: www.indymedia.org.uk/en/2006/08/349348.html (accessed 04 September 2013).

Ingham, R. (1978) 'Privacy and psychology', in J. Young (ed.), *Privacy*. Chichester: Wiley and Sons.

Johnson, C. (2013) 'Judge: Kids must wear locator chips at Texas school', *HLNtv.com*.

Jones, R. (2009) 'Checkpoint security: gateways, airports and the architecture of security', in H.O.G. K. F. Aas, and H. M. Lomell (eds), *Technologies of InSecurity: The Surveillance of Everyday Life*. Oxon: Routledge-Cavendish.

Kaliski, B.S. (no date) 'Thoughts on Radio Frequency Identification (RFID) Technology and Revelation's "Mark of the Beast"'. Available at: www.infosecwriters.com/hhworld/hh10/rfid.htm (accessed 04 September 2013).

Kellner, D. (2000) 'Multiple Literacies and Critical Pedagogies', in P. Trifonas (ed.), *Revolutionary Pedagogies*. New York: Routledge.

Koskela, H. (2003) '"Cam Era" – the contemporary urban Panopticon', *Surveillance & Society* 1 (3): 292–313.

Koskela, H. (2004) 'Webcams, TV shows and mobile phones: empowering exhibitionism', *Surveillance & Society* 2 (2/3).

Kupchik, A. (2010) *Homeroom Security: School Discipline in an Age of Fear*. New York: New York University Press.

Lehman, S., "Locator chips keep track of students in Brazil," *The Guardian*, 22 March 2012.

Lewis, T. (2003) 'The surveillance economy of post-Columbine schools', *Review of Education, Pedagogy, and Cultural Studies* 25 (4): 335–55.

Lewis, J. and Condron, S., 'Caught on a classroom spy camera – the primary school pupil who hid this girl's shoes', *Mail Online*, 11 January 2009.

Linge, M., 'Schools "spy" on fat kids', *New York Post*, 15 January 2012.

Lyon, D. (1994) *The Electronic Eye*. Cambridge: Polity Press.

Lyon, D. (2003) *Surveillance as Social Sorting*. London: Routledge.

Lyon, D. (2007) 'The search for surveillance theories', in D. Lyon (ed.), *Surveillance Studies: An Overview*. Cambridge: Polity Press.

Lyon, D. (2007) 'Resisting surveillance', in S.P. Hier and J. Greenberg (eds), *The Surveillance Studies Reader*. Berkshire: Open University Press, 368–77.

DOI: 10.1057/9781137308863

Lyon, D., ed. (2011) *Theorizing Surveillance: The Panopticon and Beyond*, 2nd edn. Oxon: Routledge.

Lyons, J., 'Hoodie hell on streets every 8 secs', *Mirror*, 4 February 2009.

Mail Online (6 November 2007) 'Hoodie Allowed to Break ASBO 16 Times Before Finally Being Sent To Prison'.

Markus, T.A. (1993) *Buildings and Power*. London: Routledge.

Mayer, M.J., (2008) 'Fact Sheet #5: School Resource Officers (SROs)' Consortium to Prevent School Violence'. www.preventschoolviolence.org (accessed 08 April 2013).

Mayer, M.J. and Leone, P.E. (1999) 'A structural analysis of school violence and disruption: implications for creating safer schools', *Education & Treatment of Children* 22 (3): 333–56.

Mccahill, M. and Finn, R. (2010) 'The social impact of surveillance in three UK schools: "angels", "devils" and "teen mums"', *Surveillance & Society* 7 (3/4): 273–89.

Mcgreal, C., 'The US schools with their own police', *The Guardian*, 09 January 2012.

Means, A. (2011) 'Securing the corporate state: education, economism, and crisis in the age of Obama', in K.J. Saltman and D.A. Gabbard (eds), *Education as Enforcement: The Militarization and Corporatization of Schools*. New York: Routledge.

Michael, J. (1994) *Privacy and Human Rights*. Hampshire: Dartmout Publishing Company Ltd.

Milne, J., 'Pupils can be tracked round town with chips', *Times Education Supplement*, 11 May 2008.

Ministry of Justice (2011) *Prison Population Projections 2011–2017 England and Wales*.

Monahan, T. (2006) 'Preface', in T. Monahan (ed.), *Surveillance and Security: Technological Politics and Power in Everyday Life*. New York: Routledge.

Monahan, T. (2006) 'The Surveillance Curriculum: Risk Management and Social Control in the Neoliberal School', in T. Monahan (ed.) *Surveillance and Security: Technological Politics and Power in Everyday Life*. New York: Routledge.

Monbiot, G., 'With threats and bribes, Gove forces schools to accept his phoney "freedom"', *The Guardian*, 4 March 2013.

Muzzati, S.L. (2004) 'Criminalising marginality and resistance: Marilyn Manson, Columbine and cultural criminology', in J. Ferrell et al. (eds), *Cultural Criminology Unleashed*. London: Glasshouse Press, 143–54.

DOI: 10.1057/9781137308863

Nolan, K. and Anyon, J. (2004) 'Learning to do time: willis' cultural reproduction model in an era of deindustrialization, globalization, and the mass incarceration of people of color', in N. Dolby, G. Dimitriadis, and P. Willis (eds), *Learning to Labor in New Times*. New York: Routledge, 114–29.

Norris, C. and Armstrong, G. (1999) *The Maximum Surveillance Society: The Rise of CCTV*. Oxford: Berg.

Norris, C. (2003) 'From the personal to the digital: CCTV, the panopticon, and the technological mediation of suspicion and social control', in D. Lyon (ed.), *Surveillance as Social Sorting. Privacy, Risk and Digital Discrimination*. London/New York: Routledge, 249–81.

Orenstein, P., 'I Tweet, therefore I am', *The New York Times*, 30 July 2010.

Parenti, C. (2000) *Lockdown America: Police and Prisons in the Age of Crisis*. New York: Verso.

Perryman, J. (2006) 'Panoptic performativity and school inspection regimes: disciplinary mechanisms and life under special measures', *Journal of Education Policy* 21 (2): 147–61.

Phillips, A. and Chamberlain, V., (2006) 'MORI Five-Year Report: An analysis of Youth Survey Data' *Youth Justice Board*, www.yjb.gov.uk.

Poster, M. (1990) *The Mode of Information: Poststructuralism and Social Context*. Chicago: The University of Chicago Press.

Qureshi, Y., 'Class CCTV comes under fire', *Manchester Evening News*, 11 March 2009.

Reno, J., Fisher, R.C., Robinson, L., Brennan, N. and Travis, J. (1999) *The Appropriate and Effective Use of Security Technologies in U.S. Schools*. Office of Justice Programs, National Institute of Justice: US Department of Justice.

Repetto, T.A. (1976) 'Crime Prevention and the Displacement Phenomenon', *Crime and Delinquency* 22: 166–177.

Robers, S., Zhang, J., Truman, J. and Snyder, T.D. (2012) *Indicators of School Crime and Safety: 2011*. National Center for Education Statistics. B.o.J. Statistics.

Robertson, A.H. (1973) *Privacy and Human Rights*. Manchester: Manchester University Press.

Roche, A.M., Pidd, K., Bywood, P., Duraisingam, V., Steenson, T., Freeman, T., and Nicholas, R. (2008) 'Drug testing in schools: evidence, impacts and alternatives', *Australian National Council on Drugs*.

Ruck, M., Harris, A., Fine, M., and Freudenberg, N. (2008) 'Youth experiences of surveillance: a cross-national analysis', in M. Flynn and D.C. Brotherton (eds), *Globalizing the Streets: Cross-Cultural Perspectives on Youth, Social Control, and Empowerment*. New York: Columbia University Press, 15–30.

Sieff, K., 'Plastic pellet incident at Va. school ends in expulsion, assault charges', *The Washington Post*, 1 February 2011.

Simon, J. (2007) *Governing Through Crime: How the War on Crime Transformed American Democracy and Created a Culture of Fear*. Oxford: Oxford University Press.

Skiba, R. and Peterson, R. (1999) 'The dark side of zero tolerance: can punishment lead to safe schools?', *Phi Delta Kappan* 80 (5): 372–82.

Skiba, R., Reynolds, C.L.R., Graham, S., Sheras, P., Conoley, J.C. and Garcia-Vazquez, E. (2006) 'Are zero tolerance policies effective in the schools? An evidentiary review and recommendations', American Psychological Association Zero Tolerance Task Force.

Stalder, F. (2002) 'Opinion: Privacy is not the antidote to surveillance', *Surveillance & Society* 1 (1): 120–24.

Staples, W.G. (2000) *Everyday Surveillance: Vigilance and Visibility in Postmodern Life*. Oxford: Rowman and Littlefield Publishers.

Steeves, V. (2012) 'Hide and seek: surveillance of young people on the internet', in D. Lyon, K.D. Haggerty, and K. Ball (eds), *The International Handbook of Surveillance Studies*. Oxon: Routledge, 352–60.

Stinchcomb, J.B., Bazemore, G. and Riestenberg, N. (2006) 'Beyond zero tolerance: restoring justice in secondary schools', *Youth Violence and Juvenile Justice* 4 (2): 123–47.

Strang, H. and Braithwaite, J. (eds) (2001), *Restorative Justice and Civil Society*. Cambridge: Cambridge University Press.

Sykes, C.J. (1999) *The End of Privacy: The Attack on Personal Rights – at Home, at Work, on-Line and in Court*. New York: St Martin's Griffin.

Taylor, E. (2010) 'Evaluating CCTV: why the findings are inconsistent, inconclusive and ultimately irrelevant', *Crime Prevention and Community Safety* 12: 209–32.

Taylor, E. (2010) 'I spy with my little eye: the use of CCTV in schools and the impact on privacy', *Sociological Review* 58 (3): 381–405.

Taylor, E. (2010) 'From finger-painting to fingerprinting: the use of biometric technology in schools', *Education Law Journal* 4: 276–88.

Taylor, E. (2011) 'UK schools, CCTV and the Data Protection Act 1998', *Journal of Education Policy* 26 (1): 1–15.
Taylor, E. (2012) 'The rise of the surveillance school', in K. Ball, K.D. Haggerty, and D. Lyon (eds), *Handbook of Surveillance Studies*. Oxon: Routledge.
Telegraph, The (2 July 2010) '"Big Brother" school lunchbox scheme abandoned'.
Telegraph, The (3 July 2010) 'Teachers able to confiscate mobile phones to control disruptive pupils'.
Tyack, D.B. (1974) *The One Best System: A History of American Urban Education*. Cambridge, MA: Harvard University Press.
US Department of Education (1994) *Fast Response Survey System, 'Survey on Advanced Telecommunications in U.S. Public Schools, K-12'*. National Center for Education Statistics.
Wacks, R. (1980) *The Protection of Privacy*. London: Sweet & Maxwell.
Wacquant, L. (2001) 'Deadly symbiosis: when ghetto and prison meet and mesh', *Punishment & Society* 3 (1): 95–134.
Wacquant, L. (2012) 'Three steps to a historical anthropology of actually existing neoliberalism', *Caderno CRH* 25: 505–18.
Waples, S., Gill, M., and Fisher, P. (2009) 'Does CCTV displace crime?', *Criminology and Criminal Justice* 9 (2): 207–24.
Warnick, R. (2010) 'Surveillance cameras in schools: an ethical analysis', *Harvard Educational Review* 77 (3): 317–43.
Weber, M. (1949/2011) *Methodology of the Social Sciences*. Ed. E.A. Shils and H.A. Finch. New Brunswick: Transaction Publishers.
Weiss, J., (2008) *Under the Radar: School Surveillance and Youth Resistance*. The City University of New York.
Weiss, J. (2011) 'Valuing youth resistance before and after public protest', *International Journal of Qualitative Studies in Education* 24 (5): 595–99.
Welland, T. (2001) 'Living in the "empire of the gaze": time, enclosure and surveillance in a theological college', *The Sociological Review* 49 (1): 117–35.
Westin, A. (1967) *Privacy and Freedom*. New York: Antheneum.
Whitaker, R. (1999) *The End of Privacy: How Total Surveillance is Becoming a Reality*. New York: New Press.
Willis, P. (1977) *Learning to Labour: How Working Class Kids Get Working Class Jobs*. Farnborough: Saxon House.

DOI: 10.1057/9781137308863

Wilson, D., Sharp, C., and Patterson, A. (2006) 'Young people and crime: findings from the 2005 offending, crime and justice survey', Home Office, *Home Office Statistical Bulletin.*

Yell, M.L. and Rotalszki, M.E. (2000) 'Searching for safe schools: legal issues in the prevention of school violence', *Journal of Emotional Behavioral Disorders* 8: 187–209.

DOI: 10.1057/9781137308863

Index

DOI: 10.1057/9781137308863

DOI: 10.1057/9781137308863

DOI: 10.1057/9781137308863

Lightning Source UK Ltd.
Milton Keynes UK
UKOW04n0600110214

226255UK00002B/17/P

9 781137 308856